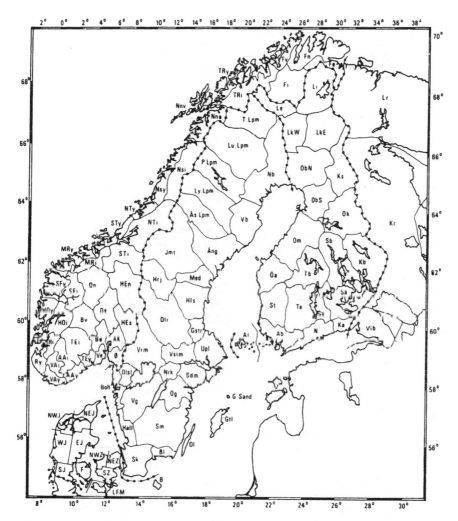

List of abbreviations for the provinces used throughout the text, on the map and in the following tables.

DENMARK

SJ	South Jutland	LFM	Lolland, Falster, Møn
EJ	East Jutland	SZ	South Zealand
WJ	West Jutland	NWZ	North West Zealand
NWJ	North West Jutland	NEZ	North East Zealand
NEJ	North East Jutland	B	Bornholm
F	Funen		

SWEDEN

Sk.	Skåne	Vrm.	Värmland
Bl.	Blekinge	Dlr.	Dalarna
Hall.	Halland	Gstr.	Gästrikland
Sm.	Småland	Hls.	Hälsingland
Øl.	Øland	Med.	Medelpad
Gtl.	Gotland	Hrj.	Härjedalen
G. Sand.	Gotska Sandön	Jmt.	Jämtland
Øg.	Østergötland	Ång.	Ångermanland
Vg.	Västergötland	Vb.	Västerbotten
Boh.	Bohuslän	Nb.	Norrbotten
Dlsl.	Dalsland	Ås. Lpm.	Åsele Lappmark
Nrk.	Närke	Ly. Lpm.	Lycksele Lappmark
Sdm.	Södermanland	P. Lpm.	Pite Lappmark
Upl.	Uppland	Lu. Lpm.	Lule Lappmark
Vstm.	Västmanland	T. Lpm.	Torne Lappmark

NORWAY

Ø	Østfold	HO	Hordaland
AK	Akershus	SF	Sogn og Fjordane
HE	Hedmark	MR	Møre og Romsdal
O	Opland	ST	Sør-Trøndelag
B	Buskerud	NT	Nord-Trøndelag
VE	Vestfold	Ns	southern Nordland
TE	Telemark	Nn	northern Nordland
AA	Aust-Agder	TR	Troms
VA	Vest-Agder	F	Finnmark
R	Rogaland		

n northern   s southern   ø eastern   v western   y outer   i inner

FINLAND

Al	Alandia	Kb	Karelia borealis
Ab	Regio aboensis	Om	Ostrobottnia media
N	Nylandia	Ok	Ostrobottnia kajanensis
Ka	Karelia australis	ObS	Ostrobottnia borealis, S part
St	Satakunta	ObN	Ostrobottnia borealis, N part
Ta	Tavastia australis	Ks	Kuusamo
Sa	Savonia australis	LkW	Lapponia kemensis, W part
Oa	Cstrobottnia australis	LkE	Lapponia kemensis, E part
Tb	Tavastia borealis	Li	Lapponia inarensis
Sb	Savonia borealis	Le	Lapponia enontekiensis

USSR

Vib   Regio Viburgensis   Kr   Karelia rossica   Lr   Lapponia rossica

FAUNA ENTOMOLOGICA SCANDINAVICA

Volume 10   1982

The Buprestidae (Coleoptera) of Fennoscandia and Denmark

by

Svatopluk Bílý

SCANDINAVIAN SCIENCE PRESS LTD.

Klampenborg . Denmark

© *Copyright*
Scandinavian Science Press Ltd. 1982

Fauna entomologica scandinavica
is edited by "Societas entomologica scandinavica"

Editorial board
Nils M. Andersen, Karl-Johan Hedqvist, Hans Kauri,
Harry Krogerus, Leif Lyneborg, Ebbe Schmidt Nielsen,
Hans Silfverberg

Managing editor
Leif Lyneborg

World list abbreviation
Fauna ent. scand.

Printed by
Vinderup Bogtrykkeri A/S
7830 Vinderup, Denmark

ISBN 87-87491-42-7
ISSN 0106-8377

Contents

Plate 1 is arranged between pp. 16 and 17
Plate 2 is arranged between pp. 32 and 33

Introduction

The family name Buprestidae was first used by Stephens (1829) and Eschscholtz (1829) and is based on the name *Buprestis* Linné, 1758. Actually, the name *Buprestis* was originally used by Linné in the 1735-edition of "Systema Naturae". The name is composed of two Greek words: bous = cattle and preto = I poison. The representatives of the family Buprestidae are, however, not poisonous, and it is obvious that some confusion between buprestids and meloids has taken place.

The Linnean genus *Buprestis* was split by Fabricius (1801) into the genera *Buprestis* and *Trachys*, and Latreille (1810) described a third genus – *Aphanisticus*. The next 12 genera of palaearctic buprestids were created by Eschscholtz (1829).

The original description of Linné's reads as follows: "Antennae filiformes, serrate longitudine thoracis. Palpi quatuor filiformes; articulo ultimo obtuso, truncato. Caput dimidium intra thoracem retractum". Originally no less than 120 species were included in the genus.

The family Buprestidae comprises about 12,000 species in the world fauna and about 1,500 species in the palaearctic fauna. There are about 200 species in Europe, 48 of which reach also northern Europe. The buprestids have an enormous range in size, shape, and coloration. The smallest species are only about 2 mm long (some *Micrasta* and *Habroloma*), while the largest species reach about 85 mm in total length *(Chrysochroa bicolor)*. Some species are narrow and elongate *(Cylindromorphus)*, other quite flat and roundish (some *Polybothris*). We can find among them brachypterous forms *(Xenorhipis)*, and species occur in the subfamily Schizopinae which resemble Dascillidae more than Buprestidae. The variability in coloration is also incredible, from entirely black species to species bearing all colours of the spectrum, from straw-yellow species to brightly metallic ones.

Also the biology of buprestids is very diverse. Most of them are wood-borers in the larval stage, but larvae of many genera undertake their development in soil (subfamily Julodinae), or they are leaf-miners (subfamily Trachyinae).

There are several taxonomic works dealing with the buprestid fauna of Europe, but they cover only West, Central, and partially East Europe, e. g. Reitter (1911), Théry (1942), Schaefer (1949), Richter (1949, 1952), Pochon (1964), Bílý (1977), and Harde (1979). Any paper has not been devoted to taxonomy, nomenclature and faunistics of the Scandinavian Buprestidae in general, except for Hansen's treatment of the family in "Danmarks Fauna" (Hansen, 1966). There are, of course, plenty of smaller papers dealing with the biology of various species, and with faunistic records. The most important of these are mentioned in the text. Krogerus (1922, 1925) published two studies of the genus *Agrilus* in Finland, but many changes have appeared in the taxonomy of this genus since that time. Much faunistic information is also to be found in Horion (1955). In the present paper is used the recent classification proposed by Cobos (1980).

9

The diagnostic characters given in the keys of higher taxa (subfamilies and tribes) have only value for the European fauna.

Acknowledgements

For the kind loan of material from various institutions and for faunistic information I am much indebted to Dr. L. Lyneborg and Mr. O. Martin of the Zoological Museum, Copenhagen; Dr. H. Silfverberg of the Zoological Museum, Helsinki; Dr. T. Kvamme of the Norwegian Forest Research Institute, and Mr. S. Lundberg of Luleå.

I am especially indebted to Dr. Leif Lyneborg for directions, advice and help in preparing the manuscript, to my friend Miss Simona Brantlová, who prepared the excellent colour plates, and to Dr. Brian Levey for linguistic correction of the manuscript.

Morphology of the adult

Head. The head is of a hypognathous type, short and vertical, with biting mouth parts, and the posterior part usually covered by the pronotum up to level of eyes. The eyes are oval, elliptical, or reniform, sometimes projecting beyond the outline of head. The labrum is transverse, incurved anteriorly. The epistome is wide, transverse, usually incurved or incised anteriorly. The frons is flat, vaulted or grooved medially, with or without a distinct pubescence. The vertex is usually vaulted, rarely flat or grooved medially.

The antennae are moderately long, eleven-segmented, usually serrate from segment 4. The basal segment is spherical or pear-shaped. Segment 2 is small, the smallest of the antennal segments, and usually spherical or slightly longer than wide. Segment 3 is cylindrical, sometimes slightly triangular. Segments 4–10 are serrate, rarely wider than long. Terminal antennal segment elliptical, trapezoidal or triangular. The antennal sensory pores are dispersed, or may be concentrated in apical or ventral sensory pits. The antennal sclerites are shallow, but sometimes bordered by a low carina.

The submentum and mentum are usually not divided into two distinct parts; the labium is small and membranous, bearing a pair of three-segmented labial palpi; paraglossae are not developed. The maxillae are well developed, with galea and lacinia membranous, and with four-segmented maxillar palpi; basal segment of maxillary palpus usually very small. The mandibles are dark, well sclerotized, robust, with a vaulted dorsal surface, and with a sharp inner edge which usually is provided with sharp teeth.

Thorax. The prothorax is relatively short, usually transverse or short cylindrical, with well developed pronotum. The pronotum is vaulted, rarely flat, very often grooved medially or depressed at posterior angles, which are more distinct than the anterior ones. In some cases the pronotum has transverse depressions. The anterior pronotal

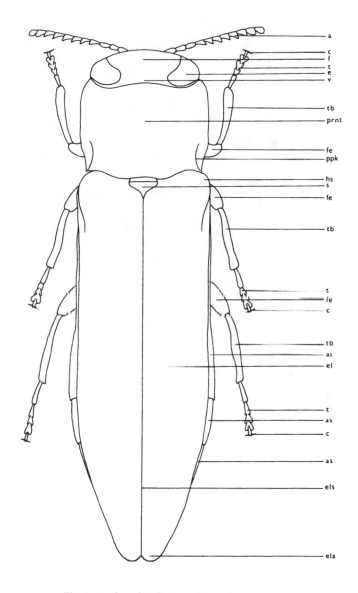

Fig. 1. *Agrilus sulcicollis* Lac., dorsal view.

a- antenna, as- abdominal segments, c- claws, e- eye, el- elytra, ela- elytral apex, els- elytral suture, f- frons, fe- femora, hs- humeral swelling, ppk- prehumeral pronotal keel, prnt- pronotum, s- scutellum, t- tarsi, tb- tibiae, v- vertex.

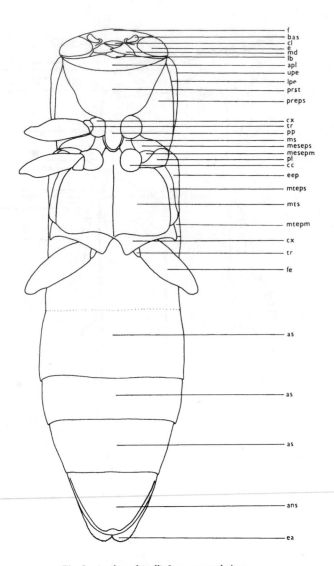

f
bas
cl
e
md
lb
apl
upe
lpe
prst
preps

cx
tr
pp
ms
meseps
mesepm
pl
cc
eep

mteps

mts

mtepm

cx

tr

fe

as

as

as

ans

ea

Fig. 2. *Agrilus sulcicollis* Lac., ventral view.

ans- anal segment, apl- anterior prosternal lobe, as- abdominal segments, bas- basal antennal segment, cc-coxal cavity, cl- clypeus, cx- coxa, e- eye, ea- elytral apex, eep- elytral epipleura, f- frons, fe- femora, lb- labium, lpe- lower pronotal edge, md- mandible, mesepm- mesepimeron, meseps-mesepisternum, ms- mesosternum, mtepm- metepimeron, mteps- metepisternum, mts- metasternum, pl- pleuron, pp- prosternal process, preps- proepisternum, prst- prosternum, tr- trochanter, upe- upper pronotal edge.

margins is straight, arched or lobate medially, while the posterior margin is usually straight or bisinuate. The lateral pronotal margins are rounded or slightly angulate, usually incurved before posterior angles or, rarely, straight. The lateral margins have one or, rarely (in *Agrilus*), two lateral carinae. The pronotal structure is very varied: smooth with a simple or umbilicate puncturation, with a transversally wrinkled structure, or with a very complex structure composed of oval or polygonal cells, punctures and various wrinkles *(Anthaxia)*. The prosternum is triangular with a straight, incurved, or arcuate, anterior margin, which in some cases has an anterior prosternal lobe *(Agrilus)*. The posterior part of prosternum forms a prosternal process which is usually long and flat, or grooved and pointed apically. The prosternal process, completely or partially, divides the mesosternum; in the former case it reaches up to metasternum. The proepisterna are large and well developed, the proepimera are very reduced, invisible.

The mesothorax is relatively small and reduced. The dorsal part (mesonotum) forms the scutellum, which is oval, semi-elliptical, cordiform, or trapezoidal. The mesosternum is composed of two parts, which are completely or partially divided by the prosternal process. The mesepisterna and mesepimera are well developed. The metathorax is large but dorsally completely covered by the elytra. The metasternum is large, usually grooved medially, with a distinct metepisternal suture. The metepisterna are large and elongate, usually somewhat widened anteriorly. The metepimera are small and reduced, usually triangular.

Abdomen. In all palaearctic species the abdomen is completely covered by the elytra. The abdomen is composed of 7 or 8 visible tergites and 5 visible sternites, the first two of which are separated by only a feeble suture. Sternites VI–VIII are invaginated and invisible. The basal abdominal sternite is anteriorly prolonged between the metacoxae and usually grooved medially. The anal (apical) segment is rounded, serrate, incurved, or notched apically. The pleurites are invisible in the majority of species. Only in the genus *Agrilus* pleurites are visible, here being separated from sternites by a well developed sternopleural suture.

Male genitalia. The external genitalia are composed of the invaginated segments IX+X, also called the auxiliary apparatus (Figs. 4, 5). Segments IX+X are very modified: tergite IX bilobate or arched, attached to tergite X which is oval or scale-shaped; sternite IX scale-shaped or slightly elongated; sternite X missing. The aedeagus is composed of a basal part, the parameres, and a phallic part. The basal part is semicylindrical, tongue-shaped or spoon-shaped, and is firmly attached to the basal part of the parameres which are elongated, usually lancet-shaped, and bear sensory apical bristles (except in some primitive groups not represented in the Scandinavian fauna). The phallic part is long and slender, lancet-shaped or cylindrical, pointed or obtuse apically, sometimes serrated apically.

Female genitalia. These are also composed of the modified urites IX and X and a membranous ovipositor (Figs. 6, 7). Tergite IX is prolonged and fork-shaped, tergite X is scale-shaped. The membranous ovipositor is usually long, apically enlarged or wing-shaped, with two short apical styli. The entire apical part of ovipositor including styli is provided with sensory bristles.

13

Wings. The membranous and hyaline hind wings are of the cantharid type with well developed radial and medial veins (Fig. 3). The elytra are free (in Scandinavian species) and strongly sclerotized, with the lateral margin slightly sinuate opposite the metacoxa. Elytral apex rounded, pointed, or serrate. Basal part of elytra with a narrow transverse basal depression and usually with well developed humeral swellings. Surface of the elytra smooth, with simple puncturation, or with a granular, scaley, or wrinkled sculpture. The elytra also have longitudinal grooves, or rows of punctures, or longitudinal keels, or, rarely, rounded depressions.

Legs. The legs are moderately long and robust. The fore and mid coxae are spherical, hind coxae transversely enlarged, reaching almost to lateral margin of the body. The trochanters are small and usually triangular. The femora are somewhat enlarged or swollen. The fore femora have a spine in *Chrysobothris,* while in *Aphanisticus* there are grooves, into which the tibiae may fit. The tibiae are straight or

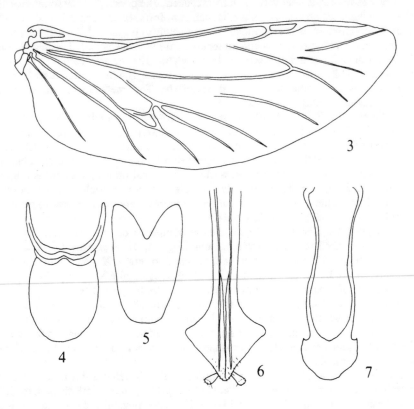

Figs. 3–7. *Anthaxia quadripunctata* (L.). – 3: right hind wing; 4: tergites IX and X of male; 5: sternite IX of male; 6: ovipositor; 7: tergites IX and X of female.

14

slightly bent, have a row of bristles on the outer margin, and are often provided with a serration on the inner margin. Exceptionally they have a hook-shaped apical spine (Fig. 80), or a medial spine, as in *Dicerca*. The tarsi are five-segmented, segments 1–4 usually being gradually larger, the terminal segment bearing a pair of claws, however, only one claw is present in *Aphanisticus*. The claws are simple, or have a more or less well developed tooth.

Sexual dimorphism. Morphological differences between the two sexes occur very frequently, and may be very pronounced. The males are usually smaller and more slender than the females. Further, differences between the sexes can be found in regard to coloration, length of antennae, shape of antennal segments, shape and width of vertex, shape of eyes, pubescence of prosternal process, shape of second and terminal abdominal segments, shape of femora and tibiae, and structure of inner margin of tibiae.

Morphology of the larva

The larvae of buprestids are apodous, dorso-ventrally flattened, almost hairless, usually with thoracic part of body very enlarged. It is possible to distinguish between four morpho-ecological types of larvae.

The first type includes larvae of the subfamilies Polycestinae, Acmaeoderinae, Chalcophorinae, Chrysobothrinae, Sphenopterinae and Buprestinae (Fig. 9). The larvae of this type have prothorax strikingly enlarged. Meso- and metathorax are somewhat narrower and very short. The ten-segmented abdomen is cylindrical, the anal segment being smooth and without thorns or teeth. The spiracles are situated on mesothorax and on abdominal segments 2–10, and are usually of a multiporous type. Only in the subfam. Acmaeoderinae are the abdominal spiracles of a uniporous type. The head is only slightly sclerotized and almost entirely hidden in the prothorax. Only the mouth parts are free. They are membraneous, except for the mandibles, epistome, hypostome and pleurostomes. The maxillae and the labium are united into a membraneous labiomaxillary complex; also the labrum and the hypopharynx are membraneous. The antennae are two- or three-segmented and short. The eyes are absent; only in some representatives of the subfam. Acmaeoderinae and Chrysobothrinae there are small rudiments of stemmata. The larvae of this morpho-ecological type develop in dying wood or bark of trees and shrubs, exceptionally in stalks of dicotyledonous herbs.

The larvae of the second morpho-ecological type (Fig. 8) belong to the subfam. Agrilinae and Cylindromorphinae. They differ from the first type by a narrower prothorax, by a longer abdomen, and primarily by the presence of sclerotized spines on the anal segment. The larvae of this morpho-ecological type undertake their development either under bark of live trees and shrubs (Agrilinae) or in grass stalks (Cylindromorphinae).

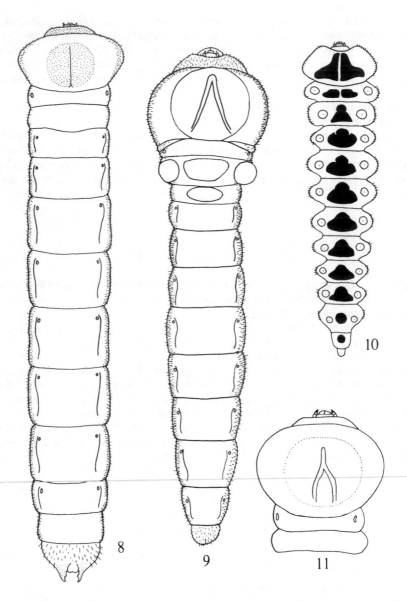

Figs. 8–11. Habitus of larvae of Buprestidae. – 8: *Agrilus biguttatus* (F.), dorsal view; 9: *Anthaxia candens* Panz., dorsal view; 10: *Trachys fragariae* Bris., dorsal view; 11: *Sphenoptera* sp., thorax in dorsal view.

16

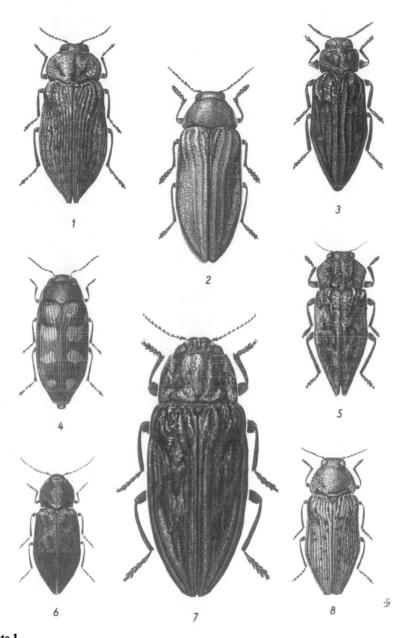

Plate 1

1. *Poecilonota v. variolosa* (Payk.), 18 mm. – 2. *Buprestis splendens* F., 20.5 mm. – 3. *Chrysobothris c. chrysostigma* (L.), 16 mm. – 4. *Buprestis o. octoguttata* L., 15 mm. – 5. *Dicerca moesta* (F.), 17 mm. – 6. *Melanophila acuminata* (DeGeer), 12.5 mm. – 7. *Chalcophora m. mariana* (L.), 30 mm. – 8. *Scintillatrix r. rutilans* (F.), 14.5 mm.

The larvae of the third morpho-ecological type (Fig. 10) are represented by the subfamily Trachyinae. They differ a great deal from the previous two groups. The body segments, including the prothorax, are rather homogenous, each segment with strongly convex lateral margins. The segments are provided with dark sclerotized plates dorsally and ventrally (in the genus *Habroloma* only dorsally). In addition, each segment, except for prothorax and anal segment, has a dorsal and a ventral pair of rounded evaginable ampullae. The anal segment is smooth and conical, without any spines. The larvae of this morpho-ecological type develop in leaf parenchyme of higher plants, where they form characteristic leaf-mines.

The fourth morpho-ecological type of buprestid larvae is represented by the subfamily Julodinae. The larvae of this type are very similar to the larvae of the subfamily Lamiinae of the Cerambycidae. They are covered by a rather long, almost lanuginose, pubescence. The prothorax is only slightly wider than the other segments and lacks the longitudinal sclerotized grooves, which are so typical for the larvae of the first and second morpho-ecological types. The body is cylindrical, robust, relatively short, and not flattened. The mandibles are ventrally prolonged into conspicuous shovel-shaped lamellae (Fig. 12), which is a quite unusual feature in buprestid larvae. The larvae of this type undertake their development in soil, feeding on roots of plants; they do not occur in Scandinavia.

The larvae of buprestids are very little known from a taxonomic point of view. Therefore the key given below is only provisional. The key to *Agrilus*-larvae is taken from Alexeev (1961). Detailed morphological descriptions of the larvae of some genera are given in the following papers: Alexeev (1960), Benoit (1964), Bílý (1971, 1972, 1974, 1975a, 1975b), Schaefer (1949), Soldatova (1973), Volkovič (1979), Yano (1952).

After the completion of this manuscript Alexeev (1981) published a key to the *Agrilus* larvae of the European part of the USSR. It contains several species not included in the following key.

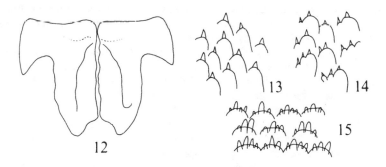

Figs. 12–15. Morphology of larvae of Buprestidae. – 12: *Julodis variolaris* Pall., mandibles in frontal view; 13–15: inner dorsal structure of proventriculus of 13: *Ptosima flavoguttata* Ill., 14: *Acmaeoderella flavofasciata* Pill., and 15: *Acmaeodera degener* Scop.

Bionomics and ecology

All buprestid beetles are heliophilous. Under the climatic conditions of Scandinavia the adults are active only in the warmest hours of the day. The only exception is *Melanophila acuminata,* which is attracted by forest fires (Burke, 1919), and it may then be numbered among the insects showing nocturnal activity.

According to the trophic relations buprestids may be divided into three groups: polyphagous, oligophagous and monophagous species. The first group is relatively small. Among typical polyphagous species we can mention only *Dicerca aenea* and *D. alni* which undertake their development in Salicaceae, Fagaceae, Betulaceae and Aceraceae, and *Melanophila acuminata* which develops in practically all Conifers. The majority of Scandinavian buprestids belongs to the oligophagous group. They undertake their development in a single plant genus or in several related genera (as most *Anthaxia, Agrilus, Buprestis* etc.). The monophagous species are also relatively rare. They develop only in one or several closely related plant species. Examples are: *Habroloma geranii* (in *Geranium sanguineum*), *Anthaxia manca* (in *Ulmus laevis* and *U. carpinifolia*), *Agrilus mendax* (in *Sorbus aucuparia* and *S. aria*), *A. integerrimus* (in *Daphne mezerum, D. laureola,* and *D. gnidium*), *A. pseudocyaneus* (in *Salix caprea* and *S. viminalis*), and *A. hyperici* (in *Hypericum perforatum* and *H. tetrapterum*).

The larvae of the Sandinavian species mainly. develop in wood of trees and shrubs (live, dying or dead), but there are several exceptions: the larva of *Agrilus hyperici* develops in roots of *Hypericum* species, the two species of *Aphanisticus* undertake their development in stalks of *Juncus* and *Carex* species, and all species of *Trachys* and *Habroloma* develop in leaf parenchyme of various herbs.

The food of adult buprestids is mostly independent of the host plants. We can observe this phenomenon e. g. in the genus *Anthaxia,* adults of which feed on flowers of Asteraceae, Ranunculaceae, Rosaceae or Daucaceae. On the contrary, adults of *Agrilus* and *Trachys* feed on leaves of their host plants.

The oviposition and fertility of buprestids is very little known. The females place their eggs singly or in small groups into ridges of bark or wood. Plochich (1969) observed that one female of *Agrilus ater* laid 16–64 eggs, the development of which lasted 16–17 days under a temperature of 19.7°–28.3°C. Jodal (1965) observed a fertility of 3–15 eggs in *Agrilus suvorovi populneus,* and the development of the eggs took 9–10 days. Arru (1961–1962) found in the same species a fertility of 2–17 eggs per female, and the development of these lasted 4–20 days. The paper by Arru is the best and most detailed study dealing with biology of any buprestid species.

The eggs are oval or spherical, white or yellowish, without microsculpture. However, the eggs of species of Trachyinae are yellow or yellowish brown with surface sculpture (Yano, 1952). Arru (1961–1962) and Lekič (1959) observed that the female covers the eggs with a secretion which hardens.

There is some uncertainty about the number of larval instars, and this certainly varies a great deal. Arru (1961–1962) found 5 instars in *Agrilus suvorovi populneus*, and Bílý (1975) 7 instars in *Anthaxia quadripunctata*. Morgan (1966) observed 6 instars in *Nascioides enysi* from New Zealand. Extremely slow development was noticed by van Dyke (1939) in the American species *Buprestis aurulenta*. The larva of this species developed over more than 30 years.

There are several more papers dealing with the biology of certain buprestid species other than those mentioned above, but these papers are not based on a study of Scandinavian species. For example, Bílý (1972) found an interesting case of developmental anomaly (prothetely) in the European species *Dicerca berolinensis* (Herbst), and Rivney (1946) described the whole development of *Capnodis tenebrionis* (L.) in Israel, and noticed a remarkable high fertility of the female of this species, 862–1696 eggs per individual!

Economic importance of buprestids

The Scandinavian buprestids are practically without any economic importance, although some of them may be potential pests of wood, e. g. *Chalcophora mariana* and some species of the genera *Dicerca* and *Buprestis*. Also *Scintillatrix rutilans* from time to time causes serious damages to old trees of *Tilia cordata* (observations from Central Europe), and *Agrilus aurichalceus* is a very important pest of cultures of *Rosa*.

Trachys minutus is usually a quite harmless species, but when occurring in dense populations it may weaken individual trees of *Salix caprea* a great deal. I observed years ago a tree of *Salix caprea* with more than 50% of the leaves damaged by larvae of *Trachys minutus*. There were from 1 to 4 larvae in each leaf. The general conclusion is, that under the climatic conditions of Scandinavia, buprestid beetles cannot be listed among serious insect pests.

Collecting, preserving and keeping

The usual method of buprestid collecting is to collect them individually from flowers, leaves, tree trunks etc. But we can collect in this way only species which visit flowers or females ovipositing on wood or in leaves. A more effective way of collecting is by sweeping the vegetation or beating of trees and shrubs. Only in this way are we able to collect large numbers of *Agrilus* or *Trachys*. Particular attention should be given to individual host plants. There are also methods to attracting buprestid beetles. For example, species developing in *Pinus* are attracted by turpentine (*Phaenops cyanea, Chalcophora mariana, Anthaxia quadripunctata, A. godeti, A. similis, Dicerca moesta, Buprestis* species etc.). Representatives of the genera *Aphanisticus, Trachys* and *Habroloma* may be collected also by sifting of detritus from under their host plants during autumn and winter.

The collecting of larvae requires a very good knowledge of host plants. Many species of buprestids can be obtained in large numbers only by rearing of adults from larvae. This method is especially effective for rearing species developing in dead wood. Species developing in live wood and herbs shows a high percentage of mortality when reared in the laboratory. The principal problem associated with rearing of adults from live wood and herbs is the appropriate humidity. Insufficient humidity is very pernicious to the larvae, and on the contrary too much humidity causes destruction of the larvae by mould and psocids.

Preserving and keeping of the adults is a very simple matter. The specimens are kept in the entomological boxes either pinned or mounted on cards. The identification of *Trachys* and *Agrilus* species is facilitated, if the specimens are mounted transversally onto the points of triangular cards. This is, because many diagnostic characters are situated on the ventral side.

Preservation of the larvae is a somewhat more complicated matter, and several methods can be used. The simplest way is to keep them in vials with 75% alcohol. However, buprestid larvae, which are soft and fatty, killed and fixed in alcohol shrink and become deformed due to a rapid dehydration. I use with very good results a solution (so-called AGO-liquid), composed of 8 portions of 96% alcohol, 5 portions of water, 1 portion of glycerine, and 1 portion of acetic acid, or a solution (Kahle's liquid), composed of 15 portions of 96% alcohol, 30 portions of water, 6 portions of 36% formaldehyde, and 4 portions of acetic acid. After 1 or 2 days, the material fixed in one of these liquids can be placed in 75% alcohol without any risk of deformation, or it may be stored in them for an unlimited time.

Faunistics

Buprestids are rather rare in Fennoscandia and Denmark. This is explained by the ecological needs of these beetles, which are usually extremely xerophilous and heliophilous. Only about 24% of the European and 3.3% of the palaearctic species occur in Fennoscandia and Denmark. The majority of the species belong to the European element (58.1%), and a smaller part to the Eurosiberian element (41.4%). Only *Agrilus aurichalceus paludicola* Krog. is known to be endemic to Fennoscandia.

The most important faunistic papers dealing with Buprestidae from Fennoscandia and Denmark are as follows: Baranowski (1980), Bjørnstad & Zachariassen (1975), Dahlgren (1964), Grill (1896), Hansen (1966), Leiler (1947), Lindroth (1960), Lundberg (1957, 1960, 1961, 1962, 1963a, 1963b, 1969, 1972, 1973, 1975, 1978, 1980a, 1980b), Lundberg, Baranowski & Nylander (1971), Huggert (1967), Lundblad (1943), Lysholm (1924, 1937), Munster (1921, 1922, 1927), Nilssen & Andersen (1977), Siebke (1875), Silfverberg (1979), Strand (1938, 1943, 1946, 1957, 1958, 1962, 1965, 1970, 1977), Strand & Hansen (1932), and Zachariassen (1972, 1977, 1979). All these publications may be found in the "Literature", and the most important of them are mentioned in the text.

European element:

Buprestis splendens
- *octopunctata*
Dicerca moesta
- *alni*
Scintillatrix rutilans
Anthaxia nitidula
- *manca*
- *similis*
- *godeti*
- *helvetica*
Agrilus angustulus
- *ater*
- *biguttatus*
- *convexicollis*
- *cyanescens*
- *guerini*
- *hyperici*
- *integerrimus*
- *laticornis*
- *mendax*
- *pseudocyaneus*
- *subauratus*
- *sulcicollis*

Aphanisticus pusillus
- *emarginatus*
Trachys troglodytes
- *scrobiculatus*
Habroloma geranii

Eurosiberian element:

Buprestis rustica
- *haemorrhoidalis*
- *novemmaculata*
Dicerca furcata
- *aenea*
Poecilonota variolosa
Melanophila acuminata
Phaenops cyanea
Anthaxia quadripunctata
Chalcophora mariana
Chrysobothris affinis
- *chrysostigma*
Agrilus aurichalceus aurichalceus
- *betuleti*
- *olivicolor*
- *pratensis*
- *suvorovi*
- *viridis*
Trachys minutus

Endemic to Fennoscandia:

Agrilus aurichalceus paludicola

Key to Scandinavian subfamilies of Buprestidae, adults

1   Elongate, oval or subcylindrical species; frons without deep
      punctiform pits above clypeal suture; pronotum at most twice
      as wide as long 2
-     Triangular or subtriangular species (Plate 2: 13); frons with
      deep punctiform pits above clypeal suture; pronotum at least
      3 times as wide as long .... **Trachyinae** (p. 86)
2 (1) Mesosternum well developed; hind coxae widest at the femo-
      ral insertion; claws without a tooth; body usually flattened and
      oval, not cylindrical; elytra at most 2.5 times as long as wide 3

21

–   Mesosternum reduced, almost indistinct; hind coxae strongly
      widening exteriorly; claws with a large tooth; body very elon-
      gate, subcylindrical and acuminate posteriorly; elytra more
      than 2.5 times as long as wide (Plate 2: 11, 16, 17) .... **Agrilinae** (p. 57)

3 (2) Fore femora with a large spine on anterior surface; radial
       cell of wing reduced; antennal sclerites separated from frons
       by a low ridge; antennal segment 3 at least twice as long as
       segment 4; eyes strongly convergent dorsally; vertex very
       narrow; each elytron with three longitudinal carinae and
       three golden green depressions (Plate 1: 3) .... **Chrysobothrinae** (p. 54)

–   Fore femora without a spine; radial cell of wing well developed;
       antennal sclerites not separated from frons by a ridge; antennal
       segment 3 shorter; vertex wider; elytra without carinae com-
       bined with golden green depressions 4

4 (3) Sensory pores on the enlarged antennal segments 5–11 are
       scattered all over the surface; scutellum extremely small,
       almost indistinct; posterior pronotal margin almost straight;
       elytra always without rows of punctures, but with shiny raised
       longitudinal areas; large species, over 20 mm long (Plate 1:7)

                                                              Chalcophorinae (p. 53)

–   Sensory pores on the enlarged antennal segments 5–11 are
       concentrated in sensory pits; scutellum large, usually trian-
       gular or pentagonal; posterior pronotal margin more or less
       bisinuate; elytra usually with longitudinal rows of punctures,
       always without shiny raised longitudinal areas; smaller species,
       less than 20 mm long .... **Buprestinae** (p. 24)

Key to European subfamilies of Buprestidae, larvae

1   Mandible large, entended ventrally into a shovel-shaped
       lamella (Fig. 12); body with a rather long and distinct pubes-
       cence, resembling that of lamiin Cerambycidae; labium
       movable, closing the mouth cavity; hypopharynx robust;
       larvae in soil; (not in Scandinavia) .... **Julodinae**

–   Mandible small, without a ventral lamella; pubescence of
       body absent or very indistinct; labium immovable; hypo-
       pharynx normal; thoracic segments usually very different
       from abdominal segments; larvae in plant tissue 2

2 (1) Abdominal segments very enlarged, laterally with rounded
       evaginable ampullae (Fig. 10); microspinules on entire body
       dark; prothorax narrower than mesothorax; pronotum,
       prosternum, and sometimes also abdominal segments, with

dark, sclerotized plates dorsally and ventrally; habitus as in Fig. 10; larvae in leaf parenchyme .... **Trachyinae** (p. 86)

– Abdominal segments without evaginable ampullae, and not extended laterally, i. e. cylindrical; microspinules, if present, white; prothorax wider than mesothorax; body segments without dark sclerotized plates; habitus as in Figs. 8 and 9; larvae in wood or bark of trees and shrubs, or in grass stalks 3

3 (2) Pronotum and prosternum without longitudinal sclerotized grooves 4

– Pronotum and prosternum with longitudinal sclerotized grooves; larvae in wood or bark of trees and shrubs 5

4 (3) Head not enlarged laterally; anal segment with two sharp, sclerotized teeth; larvae in grass stalks; (not in Scandinavia) .... **Cylindromorphinae**

– Head enlarged laterally; anal segment simple; larvae in stalks of grasses or bushes .... **Agrilinae** (partim) (p. 57)

5 (3) Anal segment well developed, with two sclerotized spines (Fig. 8) .... **Agrilinae** (partim) (p. 57)

– Anal segment very small, without spines 6

6 (5) Pronotum with a medial longitudinal groove; anterior labial margin free of bristles; mesothoracic spiracles of a multiporous type; abdominal spiracles of a uniporous type 7

– Pronotum with two grooves forming an inverted V or Y; anterior labial margin with bristles; mesothoracic and abdominal spiracles both of a multiporous type 8

7 (6) First abdominal segment shorter and narrower than abdominal segment 2; abdominal segments 2–7 1.5 times as long as wide; abdominal spiracles with peritrema; coloration bright yellow; warts on dorsal and ventral inner walls of proventriculus each with one sclerotized tooth (Fig. 13); (not in Scandinavia) .... **Polycestinae**

– First abdominal segment as long as, and wider than, segment 2; abdominal segments 2–7 wider than long; abdominal spiracles without peritrema; coloration white or cream-white; warts on dorsal and ventral inner walls of proventriculus each with several teeth (Figs. 14, 15); (not in Scandinavia) .... **Acmaeoderinae**

8 (6) Pronotal grooves forming an inverted Y (Fig. 11); antenna three-segmented; (not in Scandinavia) .... **Sphenopterinae**

– Pronotal grooves forming an inverted V (Fig. 9); antenna apparently two-segmented, as segment 3 is rudimentary 9

9 (8) Pronotum and prosternum usually with an oval field of reddish, sclerotized asperities; these fields may sometimes

be reduced to stripes along the grooves; pronotal grooves and medial prosternal groove well developed, more or less sclerotized 10

– Pronotum and prosternum without fields of asperities; pronotal and prosternal grooves only very slightly sclerotized .... **Buprestinae** (partim) (p. 24)

10 (9) Labrum lobate laterally; prothorax at most 1.5 times as wide as mesothorax 11

– Labrum not enlarged laterally; prothorax very wide and rounded, twice as wide as mesothorax; meso- and metathorax much wider than abdomen; pronotal and prosternal fields of reddish asperities always large and oval .... **Chrysobothrinae** (p. 54)

11 (10) Pronotal and prosternal fields of reddish asperities large, transversally elliptical .... **Chalcophorinae** (p. 53)

– Pronotal field of asperities more or less reduced, not oval, sometimes present in the form of stripes along the grooves; prosternal field of asperities always reduced laterally, very often found in the form of stripes along the medial groove ....
Buprestinae (partim) (p. 24)

SUBFAMILY BUPRESTINAE

Usually middle-sized species; elytra smooth, evenly punctured, or punctures forming rows; posterior pronotal margin more or less deeply bisinuate; scutellum large; sensory pores of the serrate antennal segments 5–11 concentrated in sensory pits; radial cell of wing well developed (Fig. 3).

Larva with enlarged prothorax; pronotal grooves form an inverted V (Fig. 9); antenna two-segmented; pronotum and prosternum with or without sclerotized asperities.

Key to tribes and genera of Buprestinae, adults

1 First abdominal segment prolonged at base to cover the metepimera (Fig. 16) 2

– First abdominal segment not prolonged at base to cover the metepimera (Fig. 17) 4

2 (1) Posterior pronotal margin weakly bisinuate; elytral epipleura reach apex of elytra; pronotum with granular or wrinkled sculpture (Plate 2: 9, 10, 12, 15) (Tribe Anthaxiini) *Anthaxia* Eschscholtz (p. 45)

– Posterior pronotal margin strongly bisinuate; elytral epipleura developed only on humeral section of elytra; pronotal sculpture consisting of simple punctures, which are sometimes more or less confluent (Tribe Melanophilini) 3

24

3 (2) Free margin of clypeus arcuate, flanked by sharp lateral
projections (Fig. 21); antennae enlarged from segment 4
onwards; basal segment of hind tarsi as long as combined length
of segments 2 & 3; elytron pointed apically (Plate 1:6) ....
Melanophila Eschscholtz (p. 42)
– Free margin of clypeus shallowly arcuate, flanked by low,
rounded lateral projections (Figs. 20); antennae enlarged
from segment 3 onwards; basal segment of hind tarsi shorter
than combined length of segments 2 & 3; elytron rounded
apically (Plate 2:14) .... *Phaenops* Lacordaire (p. 44)
4 (1) Mesepimera with a sharp outer anterior angle; inner and
outer margins of mesepimera almost parallel-sided (Fig. 18);
elytra distinctly striate, without smooth raised areas; (Plate
1:2, 4) (Tribe Buprestini) .... *Buprestis* Linné (p. 28)
– Mesepimera with an obtuse outer anterior angle; inner and
outer margins of mesepimera converging posteriorly (Fig. 19);
elytra with rows of punctures and with smooth raised areas (Tribe Dicercini) 5

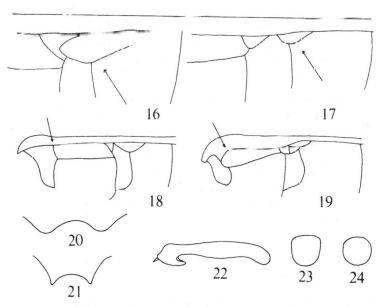

Figs. 16–24. Morphology of adults of Buprestidae. – 16: *Anthaxia hungarica* Scop., metepimeron;
17: *Buprestis rustica* L., metepimeron; 18: *Eurythyrea austriaca* Lac., mesepimeron; 19: *Dicerca berolinensis* Hbst., mesepimeron; 20: *Phaenops cyanea* (F.), clypeus; 21: *Melanophila acuminata*
(DeGeer), clypeus; 22: *Buprestis rustica* L., male fore tibia; 23: *Buprestis splendens* F., scutellum; 24:
Buprestis rustica L., scutellum.

5 (4) Elytral apex more or less elongate and acuminate; scutellum
very small, rounded or obtusely quadrangular; segments 1
and 2 of hind tarsi equally long; abdominal sternite VIII visible;
(Plate 1:5) .... *Dicerca* Eschscholtz (p. 34)
– Elytral apex not elongate or acuminate; scutellum large and
wide, usually transverse; basal segment of hind tarsi longer
than segment 2; abdominal sternite VIII invisible 6
6 (5) Metasternum and abdominal segment I deeply depressed
medially; prosternal process convex; posterior part of
elytral margins not serrate, each elytron with two small
apical spines; body dark with a bronze lustre; (Plate 1:1) ....
Poecilonota Eschscholtz (p. 40)
– Metasternum and abdominal segment I at most with a very
feeble medial depression; prosternal process flat; posterior part
of elytral margins serrate; metallic green species (Plate 1:8) ....
Scintillatrix Obenberger (p. 41)

Key to genera of Buprestinae, larvae

1 Pronotum and prosternum with reddish asperities forming
fields of varied shape; these fields may be reduced to form
narrow stripes along the grooves; prothoracic grooves well
sclerotized 2
– Pronotum and prosternum without sclerotized asperities;
prothoracic grooves less sclerotized 4
2 (1) Labrum lobate laterally; prothoracic fields of asperities
more or less reduced .... *Buprestis* Linné (p. 28)
– Labrum not lobate laterally; prothoracic fields of asperities
well developed on pronotum, oval, not reduced 3
3 (2) Pronotal and prosternal asperities small and oval, at most 1.5
times as wide as long (Fig. 29); sensory bristles on labrum
concentrated at anterior margin (Fig. 26) .... *Melanophila* Eschscholtz (p. 42)
– Pronotal and prosternal asperities transverse, at least twice
as wide as long (Fig. 30); field of sensory bristles on labrum
extends further posteriorly in midline (Fig. 27) .... *Phaenops* Lacordaire (p. 44)
4 (1) Pronotum and prosternum with a thick and coarse integu-
ment, which is matt and slightly shagreened 5
– Pronotum and prosternum with a thin, smooth and lustrous
integument; metathorax with a dorsal and ventral pair of
spherical ampullae (Fig. 9) .... *Anthaxia* Eschscholtz (p. 45)
5 (4) Pronotal grooves (especially the anterior fused part) mar-
gined by fine reddish asperities; mandible with at least three

apical teeth; the outer preapical tooth is being particularly distinct. .... *Dicerca* Eschscholtz (p. 34)

– Pronotal grooves without a margination of reddish asperities; mandible with two apical teeth; outer preapical tooth of mandible very small and indistinct 6

6 (5) Distance between the posterior apices of the inverted V formed by the pronotal grooves is about 1/3 the length of the grooves; anterior fused part of grooves not enlarged; labrum widening apically .... *Poecilonota* Eschscholtz (p. 40)

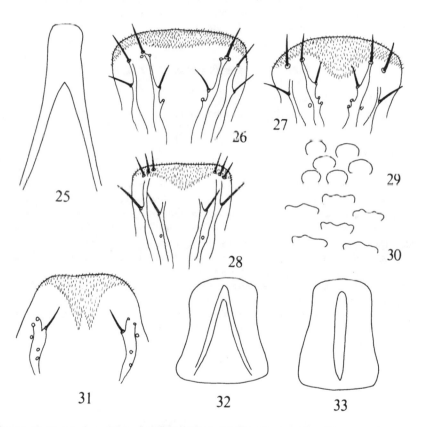

Figs. 25–33. Morphology of larvae of Buprestidae. – 25: *Scintillatrix rutilans* (F.), pronotal grooves; 26: *Melanophila picta* Pall., labrum; 27: *Phaenops cyanea* (F.), labrum; 28: *Scintillatrix rutilans* (F.), labrum; 29: *Melanophila picta* Pall., pronotal asperities; 30: *Phaenops cyanea* (F.), pronotal asperities; 31: *Melanophila acuminata* (Degeer), labium; 32: *Phaenops cyanea* (F.), pronotal plate; 33: same, prosternal plate.

–   The inverted V formed by pronotal grooves broader, the
    distance between the posterior apices about half the length of
    the grooves; anterior fused part of grooves enlarged (Fig.
    25); labrum not widening apically (Fig. 28) .... *Scintillatrix* Obenberger (p. 41)

Tribe Buprestini Kerremans, 1893

Mesepimera with a sharp outer anterior angle; inner and outer margins of mesepimera
almost parallel-sided (Fig. 18); elytra distinctly striate, without smooth raised areas.
    Larva with small reddish sclerotized asperities on pronotum and posternum.
    Only one genus which is separated in the keys above.

Genus *Buprestis* Linné, 1758, s. lat.

Buprestis Linné, 1758: 408.
    Type-species: *Buprestis octoguttata* Linné, 1758.

Ancylochira Eschscholtz, 1829: 9.
Gymnota Gistl, 1834: 10.
Anoplis Kirby, 1837: 151.

Large and oval species. Body usually dark metallic green or blue, with or without large,
pale elytral spots. Elytra with distinct striae or costae, without smooth raised areas or
apical lateral serration. Mesepimera with sharp antero-lateral angles. Scutellum small
and rounded, anterior pronotal margin not beaded. Fore tibiae of male enlarged distally or with an inner apical tooth. Eyes not projecting beyond outline of head.
    Larva with well developed sclerotized prothoracic grooves; labrum lobate laterally;
prothoracic fields of asperities more or less reduced.
    About 80 described species are known. Of these, 27 are distributed in the Palaearctic
Region, 5 also in Europe.
    The genus is divided into several subgenera: *Buprestis* s. str., *Cypriacis* Casey, *Stereosa*
Casey, *Pseudyamina* Richt. and *Orthocheira* Richt. Two of them have members in the
Scandinavian fauna.
    All our species hatch in May-June. They swarm and oviposit in June-July. The adults
feed on needles of conifers.

Key to subgenera of *Buprestis*, adults

1   Elytra with fine striae; pronotum not margined anter-
    iorly; scutellum circular (Fig. 24); fore tibiae of male with
    a large hook-shaped inner spine (Fig. 22) .... *Buprestis* Linné, s. str. (p. 30)

28

– Elytra with smooth costae; pronotum slightly margined anteriorly; scutellum more elongate (Fig. 23); fore tibiae of male only slightly enlarged distally .... *Cypriacis* Casey (p. 29)

Subgenus *Cypriacis* Casey, 1909

Cypriacis Casey, 1909: 116.
 Type-species: *Buprestis aurulenta* Linné, 1767.

Golden green species with longitudinal elytral carinae; pronotum margined anteriorly; fore tibiae of male only slightly widening distally; scutellum longer than wide.

1. *Buprestis (Cypriacis) splendens* Fabricius, 1775
Plate 1:2. Fig. 23.

Buprestis splendens Fabricius, 1775: 221.
Buprestis splendida Paykull, 1799: 229.
Buprestis pretiosa Herbst, 1801: 127.

An oval and rather flattened species. Dorsal side golden green, elytra sometimes with a blue metallic tinge. Two golden purple stripes present on elytra: one very distinct one along the elytral suture and one less distinct one along the lateral margin. Ventral side brightly golden green. Head almost flat, with long and dense white pubescence and a narrow and fine shiny carina reaching from clypeus to vertex. Puncturation of head very coarse and dense. Pronotum slightly convex, 2.1 2.2 times as wide as long. Lateral pronotal margins regularly rounded and narrowed anteriorly. Pronotum with coarse and extremely dense puncturation at lateral margins, and with sparse coarse puncturation on disc. Pronotum with a very fine and indistinct unpunctured medial line, and with a small punctiform prescutellar depression. Scutellum slightly elongate (Fig. 23). Elytra roughly punctured, with four longitudinal carinae. Apex of elytra simply rounded, without spines or serrations. ♂: epipleura with a small and sharp tooth at anterior fourth. ♀: epipleura simple, without a tooth. Length: 17.5–21.5 mm.

Distribution. Denmark: only one old record from SZ, probably an introduction. Sweden: Paykull (1799) reported the species as rare in NE Uppland; 2 specimens from Upl. in coll. NRS, and 1 specimen from Nrk., Askersund, 1895 (Ringselle leg.) in coll. GM. Not in Norway or Finland. USSR: Vib., Terijoki (= Zelenogorsk), VIII 1866. Central and South Europe, but everywhere extremely rare (Cobos, 1953).

Biology. The larva develops in old dying trunks and branches of various species of *Pinus*. Recently it has been taken in Greece, developing in dying tops of old *Pinus*-trees more than 10 m high. Larva undescribed.

Note. This species was in the past very often misidentified as the North American species *Buprestis aurulenta* Linné, 1767. This common Nearctic species was introduced to Europe, and has been recorded several times from W and SW Europe. It differs from *splendens* by the medially grooved pronotum, and by the more distinct and elevated, smooth and shiny, elytral carinae.

Subgenus *Buprestis* Linné, 1758, s. str.

Usually dark bronze, brownish or blue-green species, sometimes with yellow elytral pattern; elytra with fine striae; pronotum not margined anteriorly; scutellum small and circular.

Key to species of *Buprestis* s. str., adults

1   Elytra without spots, unicolourous brown, brown-green, green, or blue with a metallic lustre; each elytron with two or three shallow and flat depressions; 9th elytral interval somewhat elevated, forming a feeble carina 2

–   Elytra dark brown, or blue with yellow spots (Plate 2:4); elytra without shallow depressions; 9th elytral interval not elevated 3

2 (1) Lateral pronotal margins very rounded, especially in posterior half; pronotum widest at posterior third; ventral side without yellow spots; 13.0–21.0 mm .... 2. *r. rustica* Linné

–   Lateral pronotal margins rounded just before posterior angles; maximum width of pronotum at base; anterior angles of pronotum ventrally, and abdominal sternites, with very varied yellow spots; 11.0–22.0 mm .... 3. *h. haemorrhoidalis* Herbst

3 (1) First abdominal sternite flat, or with a fine shallow medial depression; body metallic blue with yellow, more or less round or oval, spots (Plate 1:4); 9.0–15.0 mm .... 4. *o. octoguttata* Linné

–   First abdominal sternite grooved medially; body dark brown with a feeble metallic tinge, and with yellow, usually irregularly shaped, spots; 12.5–21.9 mm .... 5. *n. novemmaculata* Linné

Key to species of *Buprestis* s. str., larvae

1   Prosternal groove wide, margined from the middle by several rows of reddish asperities; this margination is dilated mushroom-shaped in anterior third of the groove (Fig. 37); host plants: *Picea, Abies, Pinus* .... 2. *rustica* Linné

–   Prosternal groove narrow, margined from the middle by only one row of reddish asperities; this margination is di-

30

lated mushroom-shaped only in anterior fourth of the
groove (Figs. 36, 38) 2
2 (1) Pronotal grooves narrow, only narrowly margined by red-
dish asperities in anterior part (Fig. 35); host plant: *Pinus* .... 4. *octoguttata* Linné
-     Pronotal grooves wider, with wide margination composed
of reddish asperities (Fig. 34); host plant: *Pinus* .... 5. *novemmaculata* Linné

2. *Buprestis (Buprestis) rustica rustica* Linné, 1758
Figs. 17, 22, 24, 37.

Buprestis rustica Linné, 1758: 660.
Buprestis violacea DeGeer, 1774: 130.
Buprestis lata Sulzer, 1776: 53.

An oval and slightly convex species. Dorsal side metallic green, brownish green, blue-
green, blue, or blue with violet lustre. Ventral side copper-coloured with golden green
legs. Occasionally, clypeus, mouth parts, coxae and anal sternite have small yellow
spots. Head almost flat, with coarse puncturation. Scutellum rounded and convex (Fig.
24). Pronotum almost twice as wide as long, with coarse puncturation. Lateral pronotal
carinae almost reaching to the anterior pronotal angles. Elytra twice as long as wide
and rather flattened, with several irregular, wide and shallow, indistinct depressions.
Elytral striae deep only on humeral and sutural parts of elytra. Each elytron obliquely
truncate, apically with weak outer and inner spines. ♂: fore tibiae with a large inner
apical hook (Fig. 22); anal sternite broadly and very shallowly excavate at apex. ♀: fore
tibiae normally without a hook or spine; anal sternite broadly truncate or weakly rounded. Length: 13.0–21.0 mm.

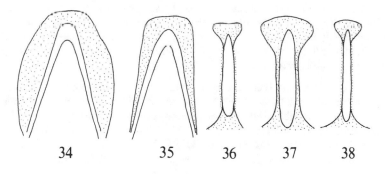

34   35   36   37   38

Figs. 34, 35. Pronotal plate of larva of 34: *Buprestis novemmaculata* L. and 35: *B. octoguttata* L.
Figs. 36–38. Prosternal plate of larva of 36: *Buprestis novemmaculata* L., 37: *B. rustica* L., and 38: *B. octoguttata* L.

31

Distribution. In Denmark probably only as an introduced species; a few records of older date from timber-yards and buildings in towns (NEZ), and also an old record from LFM. Sweden: widely distributed from Sk. to Lu. Lpm.; the most common species of the genus. Norway: in the southern districts, north to HEn; in western Norway only known from Ry. Finland: in all districts north to ObN. USSR: Vib and Kr. – Entire Europe, but not in Great Britain; throughout the USSR to the Far East, but not in Japan.

Biology. The larva develops in stumps and old dying trees of *Picea, Abies,* and *Pinus*. Also found in old house-walls of timber and in sleepers. The larval development lasts 2–3 years, pupation in May-June. The adults appear in July-September. Description of larva: Schiødte (1870), Saalas (1923), Palm (1962).

3. *Buprestis (Buprestis) haemorrhoidalis haemorrhoidalis* Herbst, 1780

Buprestis haemorrhoidalis Herbst, 1780: 97.
Buprestis punctata Fabricius 1787: 176.
Buprestis quadristigma Herbst, 1801: 177.

An elongate, rather convex species. Dorsal side dark bronze, brownish green, or bronze with a violet tinge. Ventral side and legs copper-coloured. Many yellow spots are present, namely on ventral side of body: on frons, clypeus, mouth parts, anterior angles of prosternum, and anal sternites. Head rather flat, with short white pubescence and deep, dense puncturation. Pronotum about twice as wide as long, with a fine medial unpunctured line and two smooth and shiny areas basally. Pronotal puncturation coarse and dense, especially along the lateral margins. Lateral pronotal carinae reach from posterior angles only to $^1/_3$–$^2/_3$ of pronotal length. Elytra twice as long as wide at humeral level, with indistinct and irregular, shallow depressions. Interstices between striae very flat, rather shiny. Each elytron truncate at apex, with two very fine apical spines. ♂: fore tibiae with a large apical hook on inner margin; fore femora enlarged; anal sternite truncate at apex, with two small spines. ♀: fore tibiae and fore femora normal; anal sternite shallowly bisinuate, the medial part slightly rounded. Length: 11.0–22.0 mm.

Distribution. Denmark: there are no certain records of breeding populations, and only a few records exist from EJ and NEZ in the 19th Century. Captures in buildings and timber-yards probably deal with introduced specimens. Sweden: in most districts from Sk. north to Hrj. and Nb. Norway: Ø, AK, HEs, Bø, TEy, and AAy. Finland: in nearly all districts north to Ok and ObN. USSR: Vib and Kr. – The nominate ssp. in Spain to the Urals, Italy to N. Scandinavia. Other ssp. in East Palaearctic.

Biology. The larva develops in standing, dead or dying, *Picea, Abies* and *Pinus*, with the preference for *Pinus*. Also in stumps of *Picea* and in wood laying on the ground, for instance sleepers. The development lasts 2–3 years, pupation in May-June. Adults ap-

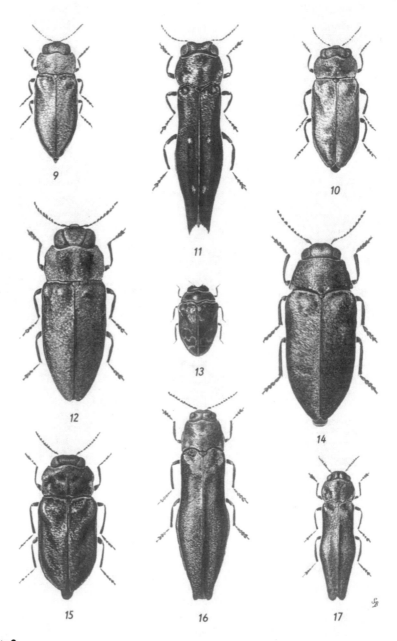

Plate 2

9. *Anthaxia n. nitidula* (L.), male, 5.5 mm. – 10. Same, female, 6 mm. – 11. *Agrilus guerini* Lac., 11 mm. – 12. *Anthaxia manca* (L.), 10 mm. – 13. *Trachys m. minutus* (L.), 3.3 mm. – 14. *Phaenops cyanea* (F.), 11 mm. – 15. *Anthaxia quadripunctata* (L.), 6.5 mm. – 16. *Agrilus subauratus* Gebl., 10 mm. – 17. *Agrilus p. pratensis* (Ratz.), 5.5 mm.

pear in June-September and can be met with on sawed wood. Description of larva: Saalas (1923), Palm (1962), Richter (1952, of ssp. *sibirica* Fleisch.).

4. *Buprestis (Buprestis) octoguttata octoguttata* Linné, 1758
Plate 1:4. Figs. 35, 38.

Buprestis octoguttata Linné, 1758: 408.
Buprestis albopunctata DeGeer, 1774: 132.

An oval and regularly convex species. Both dorsal and ventral side blue, blue-green, or blue-violet, with an extremely variable yellow pattern. Head slightly convex, with short white pubescence and simple dense puncturation. Frons with a fine smooth medial line. Pronotum rather convex, twice as wide as long, with coarse, but almost regular, puncturation; without median line or smooth raised areas. Lateral pronotal margins usually with yellow stripes. Lateral carinae of pronotum reaching from posterior angles only to middle of lateral margins. Scutellum slightly transverse. Elytra 1.8 times as long as wide, with well developed deep striae. Interstices flat, with rows of very fine punctures. Each elytron truncate apically, with two spines. ♂: fore tibiae with a deep preapical incision on inner margin; anal sternite truncate, with two lateral apical spines. ♀: fore tibiae simple; anal sternite truncate and slightly granulated on apical margin. Length: 9.0–15.5 mm.

Distribution. Not in Denmark. In Sweden north to Upl. and Dlr. Norway: rare, found in Ø, HEs, Bø, AAy, and VAy. Finland: Ab, N, Ka, Sa. USSR: Vib, Kr. – The nominate ssp. occurs all over Europe: Spain to Central Asia, and Italy to C. Scandinavia. Other ssp. in the Mediterranean area.

Biology. The larva develops in trunks and superficial roots of various species of *Pinus;* also often found in board and other pieces of wood lying on warm sandy ground. Pupal chamber in the wood. Pupation in June. Adults in June-August. Description of larva: Richter (1952), Palm (1962).

5. *Buprestis (Buprestis) novemmaculata novemmaculata* Linné, 1767
Figs. 34, 36.

Buprestis novemmaculata Linné, 1767: 662.
Elater tetrastichon Linné, 1767: 656.
Buprestis flavopunctata DeGeer, 1774: 129.
Buprestis maculata Fabricius, 1781: 275.
Buprestis flavomaculata Fabricius, 1787: 177.
Buprestis maculosa Gmelin, 1790: 1929.
Buprestis major Olivier, 1790: 32.
Buprestis octomaculata Pallas, 1806: 12.

An oval and elongate species. Dorsal side dark bronze, green-bronze, or bluish-bronze, with irregular yellow or orange spots. Ventral side dark copper-coloured, with many very variable yellow or orange spots. Head very slightly convex, with short, but dense, white pubescence and a fine dense puncturation. Vertex dark, frons yellow or orange, with irregular dark spots. Mandibles and genae usually yellow. Pronotum rather convex, 1.8–2.0 times as wide as long, roughly and densely punctured. A medial smooth and shiny line, as well as two smooth irregular lateral raised areas, well developed. Lateral margins of pronotum with yellow or orange stripes. Yellow stripe along anterior pronotal margin interrupted in the middle. Elytra 1.9–2.0 times as long as wide at humeral level, with well developed and deep striae. Interstices slightly convex, with coarse puncturation. Yellow elytral pattern very variable, rarely absent. Each elytron with three short spines apically. ♂: fore femora and fore tarsi enlarged; fore tibiae bent, with a large inner apical hook; anal sternite with a weak emargination at apex. ♀: first pair of legs normal, not enlarged and without tibial hooks; anal sternite rounded apically. Length: 12.5–21.0 mm.

Distribution. Denmark: several specimens observed c. 1830 in LFM, near Stubbekøbing; probably a natural occurrence; also found as introduced in EJ. In Sweden in some of the southern districts; Sk., Bl., Sm., Gtl., Ög., Vg., and Sdm. Norway: one uncertain record from Bø. The presence in Norway today should be verified. Finland: Ab, N, St, Ta, and Tb. USSR: Vib and Kr. – The nominate ssp. is distributed throughout Europe from Spain to Siberia.

Biology. The larva develops in wood of dead or dying trunks of various species of *Pinus*, occasionally also of *Picea excelsa*. The pupal chamber is found in the wood. Pupation takes place in June. Development lasts for at least two years, adults appearing in July-August on sawed wood of pine and on logs. Description of larva: Richter (1952), Palm (1962).

Tribe Dicercini Kerremans, 1893

Rather large, bronze or black-bronze, species; first abdominal segment not prolonged at base to cover metepimera (Fig. 17); mesepimera with an obtuse outer anterior angle; elytra with rows of punctures and smooth raised areas.

Larva characterized by pronotum and prosternum having a thick and coarse integument, which is matt and slightly shagreened; pronotal grooves well developed.

The three genera are keyed out on p. 24 ff.

Genus *Dicerca* Eschscholtz, 1829, s. lat.

Dicerca Eschscholtz, 1829: 9.
Type-species: *Buprestis acuminata* Pallas, 1782.
Latipalpis Solier, 1833: 287 (partim).
Stenuris Kirby, 1837: 154.

Large, brown or brown-bronze species with a very small scutellum. The genus is characterized by the laterally margined prosternal process, which is not grooved medially and by the antennal segment 4–11 being triangular with somewhat obtuse outer angles. Elytra usually more or less elongated apically.

Mandibles of larva with at least three apical teeth; pronotal grooves margined anteriorly by fine reddish asperities.

There are about 80 species in the world fauna, 16 species in the palaearctic fauna. In Europe 7 species, of which 4 are known to occur in Fennoscandia. The genus was divided by Gistl (1834) into two subgenera: *Dicerca* s. str. and *Argante* Gistl.

Key to subgenera of *Dicerca* s. lat., adults

1   Prosternal process flat, simply punctured; sternite I only weakly depressed; pronotum cordiform, narrow in posterior part; mid tibiae of male simple .... *Argante* Gistl (p. 35)
–   Prosternal process deeply depressed medially, with elevated margins; sternite I deeply depressed; pronotum not cordiform, only slightly narrowed posteriorly; mid tibiae of male with a tooth .... *Dicerca* Eschscholtz s. str. (p. 36)

Key to subgenera and species of *Dicerca* s. lat., larvae

1   Anterior fused section of pronotal grooves not margined anteriorly by sclerotized asperities; host plant: *Pinus* (subgenus *Argante* Gistl) .... 6. *moesta* (Fabricius)
–   Anterior fused section of pronotal grooves also margined anteriorly by sclerotized asperities (subgenus *Dicerca* s. str.) 2
2 (1) Epistome deeply incurved anteriorly; sclerotized asperities at anterior part of pronotal grooves as in Fig. 39, host plant: *Betula* .... 7. *furcata* (Thunberg)
–   Epistome less incurved anteriorly; sclerotized asperities at anterior part of pronotal grooves as in Fig. 40; hos plant: *Alnus* .... 9. *alni* (Fischer)

Subgenus *Argante* Gistl, 1834

Argante Gistl, 1834: 10.
    Type-species: *Buprestis moesta* Fabricius, 1794.
Prosternal process flat and with simple puncturation; first abdominal sternite with only a fine medial groove; pronotum cordiform, narrowed in posterior part.

6. ***Dicerca (Argante) moesta*** (Fabricius, 1793)
    Plate 1:5.

Buprestis moesta Fabricius, 1792: 206.
Buprestis quadrilineata Herbst, 1801: 104.
Dicerca divaricata J. Sahlberg, 1900: 116.

A short, matt and robust species, with coarse sculpture and puncturation. Dorsal side bronze or black-bronze, ventral side reddish-copper or copper-coloured, with purple-violet legs. Pronotum 1.75 times as wide long, subcordiform, with a deep and wide medial longitudinal depression, and with four more or less distinct longitudinal carinae. Laterobasal pronotal depressions very wide and well developed, extending anteriorly. Prosternal process without shiny and smooth lateral carinae. Elytra robust and convex, with rows of coarse and deep punctures, 1.8 times as long as wide at humeral level. Elytral apex simply narrowed, without extended elytral tips. Each elytron obliquely truncate or slightly incurved apically, with a very indistinct inner and outer tooth, or without such teeth. Black, smooth elytral raised areas small, but distinct. ♂: anal sternite slightly incurved apically. ♀: anal sternite rounded apically. Length: 12.0–19.0 mm.

Distribution. Only as introduced in Denmark (NEZ). In Sweden absent from southern districts as Sk., Bl., and Hall., but occurs in many districts from Sm. and Öl. north to T. Lpm. Norway: only Ø, AK, and AAy. Finland: Ab, N, Sa, Oa, Tb, and ObN. USSR: Vib and Kr. – Central Europe, Fennoscandia, eastwards to the Urals; also Italy incl. Sicily.

Biology. The larva develops usually in *Pinus*, but may also be found in *Picea* (Lundberg, 1957). Often in *Pinus*-trees only 1–3 m high, especially where these are growing on sun-exposed hills, but also in branches and trunks of larger trees. The development may last 3–6 years. In Sweden the pupation takes place in July. The adults hatch in late July and early August. They leave the pupal chamber and hibernate, and swarm in May, June and the beginning of July the following year. The feeding takes place on the needles of the host trees. Description of larva: Saalas (1923), Palm (1962).

Subgenus *Dicerca* Eschscholtz, 1829, s. str.

Prosternal process with a deep medial groove and raised lateral margins; first abdominal sternite deeply grooved; pronotum slightly narrowed posteriorly.

Key to species of *Dicerca* s. str., adults

1 Elytra very elongated and acuminate apically, the sutural
 margins divergent for some distance from apex; apical part
 of each elytron rounded and at most with a very small inner
 spine (Fig. 43); head without a longitudinal groove; pronotum
 deeply grooved medially, only 1.5 times as wide as long; elytra
 2.5 times as long as wide; slender species; 13.5–22.0 mm .. 7. *f. furcata* (Thunberg)
– Elytra only slightly elongated apically, the sutural margins at

36

most only slightly divergent at the extreme apex; each ely-
tron concave apically, with an inner and outer spine (Fig. 44) 2
2 (1) Robust species; elytra only twice as long as wide; elytral interstices without dark and shiny raised areas; mid tibiae of male
with only a feeble and obtuse spine (Fig. 42); 16.0–23.0 mm .... 8. *a. aenea* (Linné)
– Less robust species; elytra 2.5 times as long as wide; elytral
interstices with small, dark and shiny raised areas; mid tibiae
of male with a sharp and large spine (Fig. 41); 15.0–22.5 mm .... 9. *alni* (Fischer)

7. *Dicerca (Dicerca) furcata* (Thunberg, 1787)
Figs. 39, 43.

Buprestis acuminata Pallas, 1782: 69 (nec DeGeer)
Buprestis furcata Thunberg, 1787: 52.
Buprestis calcarata Fabricius, 1801: 188 (nec Mannerheim).

A long slender species. Entire body bronze, sometimes with a green metallic tinge, very
rarely wholly black. Frons flat, without depressions. Pronotum 1.5 times as wide as
long, distinctly grooved medially. Four well developed, smooth and shiny, longitudinal
raised areas on pronotum. Laterobasal pronotal depressions shallow and oblique.
Elytra convex and subparallel, with well developed irregular smooth black raised areas.
Elytral grooves composed of large and distinct deep punctures. Apex of elytra conspicuously elongated and acuminate, the sutural margins separate in the acuminate

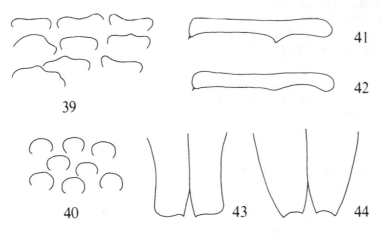

Figs. 39, 40. Pronotal asperities of larva of 39: *Dicerca furcata* (Thunb.) and 40: *D. alni* (Fisch.).
Figs. 41, 42. Male mid tibia of 41: *Dicerca alni* (Fisch.) and 42: *D. aenea* (L.).
Figs. 43, 44. Apex of elytra of 43: *Dicerca furcata* (Thunb.) and 44: *D. aenea* (L.).

37

part. Each elytron with only a small inner spine. Outer angles of elytral apex obtuse or rounded, spineless. ♂: mid tibiae with a sharp needle-shaped inner spine at anterior third; anal segment with a deep trapezoid incision apically. ♀: mid tibiae simple and smooth; anal sternite with two deep incisions apically, forming three spines. Length: 13.5–22.0 mm.

Distribution. Not in Denmark. Sweden: as in the former species apparently absent in the southern districts as Sk., Bl., and Hall., recorded from a number of districts and rather frequent in northern Sweden, from Sm. north to Lu. Lpm. Norway: only records from AK, Bø, and TEy. Finland: most districts from Ab and N in the south to Ok and LkW in the north; most records are from the 19th Century. USSR: Vib and Kr. – Throughout palaearctic Eurasia: Spain to the Far East of the USSR and China; Italy to N. Scandinavia. Ssp. *aino* Lewis, 1893 in Japan.

Biology. The larva develops in dead and dying trunks and branches of species of *Betula;* it has in Sweden also been reared from *Populus tremula.* It bores usually in hard wood without bark. The developmental period lasts 3 years in Sweden. Pupation in June, adults appearing in August, and after hibernation they swarm in the following spring.

8. *Dicerca (Dicerca) aenea aenea* (Linné, 1761)
Figs. 42, 44.

Buprestis aenea Linné, 1761: 213.
Buprestis austriaca Schrank, 1781: 195.
Buprestis oxyptera Pallas, 1782: 10.
Buprestis cuprea Rossi, 1790: 184.
Buprestis reticulata Fabricius, 1794: 451.
Buprestis subrugosa Paykull, 1799: 218.
Buprestis carniolica Fabricius, 1801: 189 (nec Herbst).
Dicerca scabrosa Mannerheim, 1837: 54.

A large robust and very convex species: Dorsal side bronze or rarely black, with a slight metallic lustre; ventral side with a coppery lustre. Pronotal medial groove indistinct, pronotum without smooth longitudinal raised areas. Elytra very robust, 1.8–2.0 times as long as wide at humeral level. Apical part of elytra elongated, slightly acuminate, with the sutural margins somewhat separate. Each elytron with a smaller inner and a larger outer spine (Fig. 44). Smooth, black raised areas of elytra very small, developed only along the suture. ♂: metasternum grooved medially; anal sternite with shallow notch and two spines apically; mid tibiae with low inner tooth at middle of inner side (Fig. 42). ♀: metasternum flat without any groove; anal sternite with two shallow incisions, forming three spines; mid tibiae simple. Length: 16.0–23.0 mm.

Distribution. Not in Denmark. Sweden: Sk., 2 specimens c. 1850 (Mus. Lund). Norway: several records in AK (Oslo) from the middle of the last century; and one uncertain record from Bø. The presence today should be verified. Not in Finland, nor in the Karelian part of the USSR. – Portugal to Central Asia, Algeria to S. Scandinavia; also known from Baltic region of the USSR: Estonia, Lithuania, Latvia. Other ssp. occur in the Middle East and China.

Biology. The larva develops in old, dead, or dying trunks, or thick branches of species of *Populus* and *Salix,* and is usually found 3–5 cm under the bark. The development lasts at least 4 years. The adults are active only on hot sunny days in May–August, sitting and swarming on the sun-exposed side of the trees. Description of larva: Perris (1877), Richter (1952).

9. *Dicerca (Dicerca) alni* (Fischer, 1823).
   Figs. 40, 41.

Buprestis mariana DeGeer, 1774: 128 (nec Linné).
Buprestis berolinensis Paykull, 1799: 217 (nec Herbst).
Buprestis alni Fischer, 1823: 191.
Buprestis fagi Gory & Laporte de Castelnau, 1841: 103.
Dicerca calcarata Mannerheim, 1837: 55 (nec Fabricius).
Lampra compressa Gernet, 1867: 17.

A large, elongate and rather convex species. Dorsal side bronze or greenish bronze, ventral side copper-red, or rarely whole body black-bronze. Pronotum 1.5 times as wide as long, with a slight medial longitudinal groove and well developed wide and shallow, laterobasal depressions. Smooth and shiny pronotal raised areas more or less developed. Elytra subparallel, 2.2 times as long as wide at humeral level and acuminate in posterior third. Small, elongate, smooth raised areas of elytra, shiny and well developed. Elytral apex elongated and caudiform, sutural margins slightly separated apically. Each elytron with two sharp apical spines of equal size. ♂: metasternum distinctly grooved medially; mid tibiae with a large and right-angled tooth at middle (Fig. 41); anal sternite with a deep right-angled notch. ♀: metasternum without a medial groove; mid tibiae simple, without a tooth; anal sternite with two shallow incisions apically, forming three spines. Length: 15.0–22.5 mm.

Distribution. Not in Denmark or Norway. In Sweden only recorded from Sk., Sm., Öl., Sdm., Upl., Vstm., and Hls.; in Finland from Al, Ab, Ta, Sa, Tb, and Om; also Vib and Kr in the USSR. – Throughout Europe; also Algeria and Tunisia.

Biology. The larva undertakes its development in the sun-exposed section on dead trunks of *Alnus* species. Richter (1952) and Schaefer (1949) also mentioned *Corylus avellana* and *Juglans regia.* Lundberg (1961) found the species breeding in tree-tops of *Tilia* on an island in Mälaren in Sweden. The larva bores its tunnel, which is situated deep in the wood, parallel with the trunk-axis. The pupal chamber is, however, situated immediately under the bark. Pupation takes place in July, adults appearing in August, and again in April after the hibernation. Description of larva: Perris (1877).

Genus *Poecilonota* Eschscholtz, 1829

Poecilonota Eschscholtz, 1829: 9.
  Type-species: *Buprestis variolosa* Paykull, 1799.

Relatively large and flattened species. Dorsal side dark bronze, with numerous small shiny black raised areas. Elytral striae deep and complete. Mesepimera with obtuse or rounded antero-lateral angles. Sensory pores of antenna concentrated in pits on the ventral side of the segments. Prosternum without an anterior lobe. Apical part of elytra not elongated, without lateral serration. Scutellum wider than long, basal segment of hind tarsus longer than segment 2. Metasternum and basal part of abdomen grooved medially, prosternal process convex. Each elytron with two apical spines.

Labrum of larva widening apically; distance between the posterior apices of the inverted V formed by the pronotal grooves about 1/3 as long as the length of the grooves; anterior fused part of grooves not enlarged.

There are 12 described species in the world fauna. Three species are known to occur in the Palaearctic Region, one of which occurs in Europe.

10. *Poecilonata variolosa variolosa* (Paykull, 1799).
    Plate 1:1.

Buprestis tenebrionis Schaeffer, 1766: pl. II, f. 1 (nec Linné).
Buprestis rustica Herbst, 1787: 174 (nec Linné).
Buprestis plebeia Olivier, 1790: 89 (nec Fabricius).
Buprestis variolosa Paykull, 1799: 219.
Buprestis conspersa Gyllenhal, 1808: 441.
Poecilonota aspersa Rosenhauer, 1856: 135.

An oval flattened species, which is somewhat broadened posteriorly. Dorsal side dark bronze with small golden elytral spots, ventral side bronze-copper or reddish copper. Pronotum flattened, 1.7 times as wide as long, with rounded or angulate lateral sides. Pronotum with a feeble, median, longitudinal smooth keel, and with shallow oblique laterobasal depressions. Also present are several small and shiny raised areas on the disc of pronotum, especially on its anterior part. Scutellum trapezoidal and wide. Elytra 1.8 times as long as wide, the maximum width at posterior third. Elytral striae deep and well developed, more shallow laterally. Each elytron truncate apically, with small outer and inner spines, elytral margins not serrate posteriorly. First abdominal segment slightly grooved medially. ♂: anal sternite with a deep semicircular incision apically. ♀: anal sternite only very slightly incised apically. Length: 12.0–20.0 mm.

Distribution. Not in Denmark. In Sweden from Sk. north to Dlr. and Nb. Norway: Ø, AK, HEs, HEn, Bø, TEy, TEi, AAy. Finland: Ab, N, St, Ta, Sa, Tb, Sb, Om. USSR: Vib and Kr. – The nominate ssp. ranges from Spain to lake Baikal, and from Greece to N. Scandinavia. Other ssp. in E. Palaearctic and NW Africa + S Spain.

Biology. The larva undertakes its development in standing, live or dying, trunks or thick branches of *Populus* species, preferably *P. tremula*. It is usually found in the bark, but also in dead wood. Pupation takes place in spring, usually in the bark. The adults swarm in June–July. Description of larva: Richter (1952), Palm (1962), Alexeev & Zykov (1979).

Genus *Scintillatrix* Obenberger, 1955

Scintillatrix Obenberger, 1955: 41.
   Type-species: *Buprestis rutilans* Fabricius, 1777.
Lampra Lacordaire, 1835: 595 (praeocc. Hübner, 1821, in Lepidoptera).

Relatively large and flattened, brightly metallic species, the elytra with numerous small black raised areas. This genus was separated from *Poecilonota* Eschsch. by Lacordaire (1835), and differs from *Poecilonota* by the metallic, usually golden green, coloration, the flat prosternal process and the sharply serrate apical margins of elytra, and by having metasternum and base of abdomen ungrooved medially.

   Larva of *S. rutilans* (F.) have the pronotal grooves (Fig. 25) narrow, the posterior apices of the grooves being comparatively widely separated; anterior fused part of grooves enlarged. Labrum (Fig. 28) comparatively narrow; field of sensory bristles wider in middle than laterally. The fine, sclerotized asperities between V-shaped pronotal grooves are 2–3 times as long as wide at base. Host-plant: *Tilia cordata*.

   There are more than 60 species in the world fauna. Most of them are found in the Indomalayan area. The palaearctic fauna comprises about 30 species, 6 of which are also known to occur in Europe.

   Kerremans (1900) described the genus *Ovalisia* from the Solomon Islands. This name was used by some authors as a younger synonym of *Lampra* Lac. (e. g. Bílý, 1977). However, Hellrigl (in pers. comm.) has studied the Kerremans type, and is of the opinion that *Ovalisia* represents a distinct oriental genus. Schaefer (1949) erected the subgenus *Palmar* for species possessing large elytral spots; Hellrigl (1972) made this subgenus a separate genus, therefore the name *Scintillatrix* Obenberger becomes the available replacement name for *Lampra* Lacordaire, 1835.

11. *Scintillatrix rutilans* (Fabricius, 1777)
    Plate 1:8. Figs. 25, 28.

Buprestis rutilans Fabricius, 1777: 235.
Buprestis aeruginosa Herbst, 1780: 91.
Buprestis rustica Schrank, 1781: 194 (nec Linné).
Buprestis fastuosa Jacquin, 1781: 385.

An oval and moderately convex species. Dorsal side brightly golden green, rarely blue-green, with a wide orange stripe along the pronotal and elytral margins. Pronotum with an indistinct black medial line, and with several very indistinct and small black spots. Elytra with many small shiny, black, smooth, raised areas. Ventral side golden green, abdominal segments with a slight orange tinge. Head relatively small; frons with small, irregular and smooth, metallic raised areas. Antennae short, reaching only to middle of lateral pronotal margins. Pronotum with coarse puncturation, about twice as wide as long. Maximum width of pronotum just before posterior angles. Laterobasal pronotal depressions large and shallow. Elytra 2.0–2.1 times as long as wide at humeral level, maximum elytral width at the second third. Apical part of elytra with irregular serration. Elytral striae deep, the interstices between them coarsely punctured and with irregular small, black, smooth, raised areas. ♂: apical margin of anal sternite with a rather deep incision between two sharp spines. ♀: anal sternite slightly incurved apically. Length: 11.0–15.0 mm.

Distribution. Not in Denmark or Sweden; Norway: AK and TEy; Finland: N (Kyrkslätt) and Sa (Savonlinna). – Spain to Causasus, Sicily to Scandinavia.

Biology. The larva develops under the bark of trunks and thick branches of various species of *Tilia;* according to some authors also in *Fagus sylvatica* (Hellrigl, 1978). Larval development lasts 1–3 years. Pupation occurs in May or June, adults appearing in May–August. Description of larva: Taschenberg (1879), Alexeev (1979).

Tribe Melanophilini Bedel, 1921

Middle-sized species; posterior pronotal margin deeply bisinuate; epipleura developed only at humeral section of elytra; pronotal structure consisting of simple punctures, which may be more or less confluent.
    Labrum of larva not lobate laterally; fields of sclerotized asperities well developed, pronotal field oval, not reduced.
    Two genera in our area. They are keyed out on p. 25 and p. 26.

Genus *Melanophila* Eschscholtz, 1829

Melanophila Eschscholtz, 1829: 9.
    Type-species: *Buprestis acuminata* DeGeer, 1774.
Trachypteris Kirby, 1837: 158.
Oxypteris Kirby, 1837: 160.
Apatura Gory & Laporte de Castelnau, 1841: 1.
Diana Gory & Laporte de Castelnau, 1841: 155.

Medium-sized, flattened species. Body black, sometimes with a metallic lustre, or dark with a bronze lustre, and with yellow elytral pattern. Metepimera partially covered by a lobe from the first abdominal segment. Posterior pronotal margin deeply bisinuate. Epipleura developed only at subhumeral parts of elytra. Pronotal structure consisting only of simple punctures or short simple transverse wrinkles. Clypeus with two sharp points bordering anterior clypeal emargination (Fig. 21). Antennae serrate from segment 4 towards. The basal segment of hind tarsus as long as segment 2 & 3 together. Each elytron with an apical spine, or pointed apically.

Larva with pronotal and prosternal sclerotized asperities small and oval, at most 1.5 times as wide as long (Fig. 29); sensory bristles on labrum concentrated to anterior margin (Fig. 26).

There are about 70 species in the world fauna; most of them are found in the Neartic and Neotropical Regions. Only 4 species occur in the Palaearctic Region; 3 of these are known to occur in Europe.

Some authors (e. g. Richter, 1949) divide the genus into two subgenera: *Melanophila* s. str. (with pointed elytral apex) and *Trachypteris* Kirby (with rounded and serrate elytral apex). Our species belong in the first mentioned subgenus.

12. *Melanophila (Melanophila) acuminata* (DeGeer, 1774)
Plate 1: 6. Figs. 21, 31.

Buprestis acuminata DeGeer, 1774: 133.
Buprestis acuta Gmelin, 1788: 1939.
Buprestis appendiculata Fabricius, 1792: 210.
Buprestis morio Fabricius, 1792: 230 (nec Herbst).
Buprestis longipes Say, 1823: 164.
Melanophila immaculata Mannerheim, 1837: 70.
Anthaxia pecchiolii Gory & Laporte de Castelnau, 1841: 33.
Melanophila assimilis Le Conte, 1852: 227.
Melanophila rugata Le Conte, 1857: 7.
Melanophila anthaxoides Marquet, 1869: 368.
Melanophila obscurata Lewis, 1893: 331.

A large, black and slightly shiny species. Entire dorsal surface without hairs; ventral surface with white, and legs with black, pubescence. Frons flat, with two small and shallow depressions; vertex 1.4 times as wide as width of an eye. Frons with a dense network of small polygonal cells; vertex with simple puncturation. Pronotum 1.3–1.4 times as wide as long, with small polygonal cells laterally, and with a simple puncturation on disc. Lateral pronotal margins rounded, almost straight before posterior angles. Medial pronotal groove very fine, but distinct. Posterior pronotal angles right-angled. Scutellum very small, longer than wide. Elytra much wider than pronotum, 2.0 times as long as wide. The widest part of elytra at apical fourth. Apical part of elytra slightly serrate, each elytron sharply pointed. Elytra with very fine granular sculpture. Invariant. ♂: mid tibiae slightly serrate on inner margin; anal sternite with a deep apical incision; prosternal pubescence denser. ♀: mid tibiae simple; anal sternite with a shallow apical incision; prosternal pubescence shorter and more sparse. Length: 6.0–13.0 mm.

Distribution. Denmark: only one record in this century, B, Rønne, 1924. This specimen was taken on the beach and is probably a vagrant; also some scattered captures (EJ, LFM, NEZ) from the 19th Century. Sweden: widespread from Sk. to T. Lpm. Norway: scattered records from the south (AK) to the very north (Fø), but not known from the western parts. Also in Finland recorded from nearly all districts. USSR: Vib, Kr, and Lr. – Very widely distributed: Algeria, whole Europe (incl. Great Britain), Caucasus, Siberia, Mongolia, the Far East, Canada, the United States, Cuba, Haiti (probably introduced to Cuba and Haiti). The species of buprestid with the widest distribution.

Biology. The larva undertakes its development under the bark, especially of the stump, of all kinds of conifers, but can also be found on birch *(Betula)*. The development lasts 2–3 years, pupation taking place in the wood. The adults are attracted by the infrared radiation caused by wood-fires, and the females oviposit preferably in trees damaged by fire. The adult beetles occur in May–September. Description of larva: Palm (1949), Soldatova (1969), Benoit (1966).

Genus *Phaenops* Lacordaire, 1857

Phaenops Lacordaire, 1857: 47.
   Type-species: *Buprestis cyanea* Fabricius, 1775.

Medium-sized, flattened and oval species. Body usually metallic blue or green, rarely black, sometimes with spotted elytra. According to some authors it is better regarded as a subgenus of *Melanophila* Eschscholtz (e. g. Obenberger, 1930). It differs from *Melanophila* by having very obtuse clypeal points (Fig. 20), bordering a shallow or indistinct anterior clypeal emargination, by having the antennae serrate from the third segment onwards, the basal segment of hind tarsi is shorter than segments 2 & 3 together, and the elytra rounded apically.
   Larva with pronotal and prosternal sclerotized asperities transverse, at least twice as wide as long (Fig. 30); field of sensory bristles on labrum extends further prosteriorly in midline (Fig. 27).
   There are 15 species in the world fauna, 13 of which are known to occur in the Palaearctic Region; 6 in Europe.

13. *Phaenops cyanea* (Fabricius, 1775)
   Plate 2:14. Figs. 20, 27, 30, 32, 33.

Buprestis cyanea Fabricius, 1775: 216.
Buprestis chalybaea Villers, 1789: 339.
Buprestis tarda Fabricius, 1794: 209.
Buprestis clypeata Paykull, 1799: 223.
Phaenops sibirica Pic, 1918: 1.

A large oval species without dorsal pubescence. Entire body dark blue, sometimes with darker elytra, or whole body blue-black or, very rarely, black. Ventral side with rather long white pubescence. Frons regularly convex, vertex 1.8–2.0 times as wide as width of an eye. Head with coarse, deep and simple puncturation. Pronotum almost conical, 1.4 times as wide at base as long, with a fine prescutellar depression. Sculpture of pronotum consisting of simple deep punctures, which form short transverse wrinkles on the prescutellar part. Posterior pronotal angles obtuse. Scutellum very small, rounded. Elytra oval, 1.6–1.7 times as long as wide, with fine preapical serrations. Elytra not pointed, each elytron separately rounded. Elytral structure consisting of coarse irregular punctures and transverse wrinkles. ♂: mid tibiae, and especially hind tibiae, serrate on inner margin; prosternal pubescence longer; anal sternite truncate. ♀: mid and hind tibiae simple; prosternal pubescence shorter; anal sternite with small apical depression, its apical margin slightly incurved. Length: 6.6–12.4 mm.

Distribution. Denmark: introduced (NEZ). Sweden: in most districts from Sk. in the south to Lu. Lpm. in the north. Norway: only in the south, records from AK, HEn, Bø, VE, TEy, TEi, and AAy. Finland: from Ab and N in the south to Sb and Kb in the north. USSR: Vib and Kr. – Ranges through most of Europe to Siberia and Mongolia; also Algeria and Morocco. Not native in Great Britain.

Biology. The larva undertakes its development in and under the bark of branches and trunks of various *Pinus* species. The development lasts 1–2 years, and the adult occur in May July. They feed on the host trees, attacking the needles of the terminal shoots. The species is attracted to pines which are damaged by fire. Description of larva: Perris (1854, 1877), Richter (1949), Alexeev (1964).

Tribe Anthaxiini Castelnau et Gory, 1839

Middle- to small-sized species; metepimeron covered by a prolongation of first abdominal sternite (Fig. 16); posterior pronotal margin shallow bisinuate; epipleura reach apex of elytra; pronotum with granular or wrinkled structure.
Larva: pronotum and prosternum without sclerotized asperities; prothoracic grooves less sclerotized.
Our single genus is keyed out on p. 24 and p. 26.

Genus *Anthaxia* Eschscholtz, 1829, s. lat.

Anthaxia Eschscholtz, 1829: 9.
Type-species: *Buprestis fulgurans* Schrank, 1789.

Small, flattened or subcylindrical species. Coloration very variable: from black to metallic blue or green, very often with a more or less complex elytral and pronotal pattern. Metepimeron partially covered by the basal prolongation of the first abdominal sternite (Fig. 16). Prosterior pronotal margin straight, or almost so. Elytral epipleura

45

complete, reaching from the humeral area to the apex of the elytra. Pronotal structure complex, consisting of a network of cells or wrinkles, with or without umbilicate punctures etc., never consisting of simple punctures.

Larva: pronotal and prosternal plates with a thin, smooth, shiny integument; metathorax with a dorsal and ventral pair of spherical ampullae (Fig. 9).

A very large genus comprising about 600 species all over the world. About 170 species are distributed in the Palaearctic Region; 60 of these are known to occur in Europe.

Richter (1949) divided the genus into three separate genera and 13 subgenera, but his division breaks down, if the world fauna is considered. I have worked on this genus for many years, and in my opinion the genus should be divided into 7 subgenera: *Anthaxia* s. str., *Haplanthaxia* Reitter, *Melanthaxia* Richter, *Cratomerus* Solier, *Agrilaxia* Kerremans, *Paracuris* Obenberger, and *Cylindrophora* Solier. Two of these are represented in the Scandinavian fauna.

Key to subgenera of *Anthaxia* s. lat., adults

1 Black and dark bronze species; body rather wide and sub-parallel; development in coniferous trees .... *Melanthaxia* Richter (p. 49)
– Rather flattened, bi- or multicolorous species; development in deciduous trees .... *Anthaxia* Eschscholtz, s. str. (p. 47)

Key to species of *Anthaxia* s. lat., larvae

1 Prosternal groove well developed, not shortened; inner structure of proventriculus consisting of short and thick, pale spines, which are at most 3–4 times as long as wide (Fig. 46) 2
– Prosternal groove less developed, only its middle section is distinctly sclerotized; inner structure of proventriculus composed of long and dense, almost bristle-like, orange spines, which are 8–10 times as long as wide as base (Fig. 45); labrum with a sensory area at the base of the inner sensory bristle (Fig. 47); host plant: *Ulmus* .... *manca* (Linné)
2 (1) Pronotal grooves regularly pigmented; inner structure of proventriculus composed of simple and doubled spines (Fig. 46); host plants: conifers .... 15. *quadripunctata* (Linné)
– Pronotal grooves irregularly pigmented: in mosaic, or only at apex 3
3 (2) Pronotal grooves with mosaic pigmentation in middle only; prosternal groove not pigmented; host plant: *Pinus* .. 16. *godeti* Castelnau et Gory
– Pronotal grooves pigmented in posterior half; anterior confluent part of pronotal grooves very enlarged; host plants: fruit trees .... 14. *nitidula* (Linné)

46

Subgenus *Anthaxia* Eschscholtz, 1829, s. str.

Key to species of *Anthaxia* s. str., adults

1　　Large species, 7.0–11.0 mm; elytra brown; pronotum orange
　　with two large black longitudinal stripes; very rarely prono-
　　tum has a green tinge; vertex narrow, about half as wide as
　　diameter of an eye; male with innerface of tibiae serrate; en-
　　tire body with a long white pubescence .... *manca* (Linné)
–　　Small species, 4.2–7.2 mm; entire body metallic golden green
　　with golden pronotum (♂), or elytra blue-green or blue and
　　pronotum orange (♀); vertex about as wide as the width of an
　　eye; male hind tibiae simple; only head and underside with a
　　short white pubescence .... 14. *n. nitidula* (Linné)

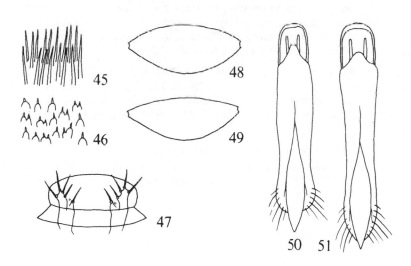

Figs. 45, 46. Inner dorsal structure of proventriculus of larva of 45: *Anthaxia manca* (F.) and 46: *A. quadripunctata* (L.).

Fig. 47. Labium of larva of *Anthaxia manca* (F.).

Figs. 48, 49. Body profile of adult of 48: *Anthaxia godeti* Cast. et Gory and 49: *A. quadripunctata* (L.).

Figs. 50, 51. Aedeagus of 50: *Anthaxia godeti* Cast. et Gory and 51: *A. quadripunctata* (L.).

14. *Anthaxia (Anthaxia) nitidula nitidula* (Linné, 1758)
    Plate 2:9, 10.

Buprestis nitidula Linné, 1758: 410.
Buprestis laeta Schaeffer, 1766: 67.
Buprestis styria Voet, 1806: 95.
Anthaxia grabowskii Obenberger, 1912: 7.

A small, subparallel and slightly convex species. Body glabrous, frons with a very short white pubescence. Entire body golden green or blue-green, pronotum sometimes golden yellow or orange (♀). Frons flat or very slightly grooved, vertex 1.0 (♂) or 1.4 (♀) times as wide as width of an eye. Frons with a network of small, densely set, polygonal cells, which have indistinct central grains. Pronotum almost twice as wide as long, with rounded lateral margins; the maximum width of pronotum occurs at the anterior third. Pronotum has two wide and shallow laterobasal depressions and a very fine medial groove. The pronotal sculpture consists laterally of polygonal cells provided with small central grains, and medially of irregular transverse wrinkles. Anterior pronotal margin with a protuberant medial lobe. Scutellum almost semielliptical, slightly vaulted. Elytra subparallel, slightly convex, 2.0–2.1 times as long as wide measured at humeral level, with fine lateral serrations at the very apex. Sculpture of elytra very fine, shiny, with sparse irregular puncturation on disc. ♂ (Plate 2:9): golden green, sometimes with golden yellow pronotum, or blue-green with green pronotum; mid and hind tibiae indistinctly serrate on inner margin; ♀ (Plate 2:10): blue-green, or blue with orange pronotum; mid and hind tibiae simple. Length: 4.2–7.2 mm.

Distribution. According to Richter (1949) in Finland, Norway, Sweden, and Denmark. I have studied only one speciemen labelled »Sweden«. – Baltic region of the USSR: Estonia, Lithuania, Latvia. The nominate ssp. ranges from Portugal and Great Britain to Central Europe, and from North Africa and South Europe to South Scandinavia.

Biology. The larva develops under the bark of trunks and twigs of various fruit trees. The development lasts 1–2 years. The pupa or the last instar larva hibernates in a pupal chamber. The adults appear in May–July, and can be found on various flowers. Description of larva: Bílý (1975).

Anthaxia manca (Linné, 1767)
    Plate 2:12. Figs. 45, 47.

Buprestis manca Linné, 1767: 1067.
Buprestis bistriata Fabricius, 1775: 22.
Buprestis elegantula Schrank, 1781: 367.
Cucujus rubinus Fourcroy, 1785: 33.
Anthaxia elongatula Kerremans, 1892: 123.
Anthaxia mancatula Abeille de Perrin, 1900: 27.

48

A large flat species with long white pubescence. Elytra dark brown, with a slight violet tinge; vertex and pronotum orange, or rarely golden green; frons and two longitudinal pronotal stripes black. Ventral side, and sometimes subhumeral parts of elytral margins, copper-coloured. Exceptionally, entire body with a greenish tinge. Frons flat, with very long white pubescence. Vertex very narrow, 0.5–0.7 times as wide as width of an eye. Sculpture of the head consists of small and densely set, roundish cells with small sharp central grains. Pronotum flat, 1.8–1.9 times as wide as long, with shallow and indistinct laterobasal depressions. Medial pronotal groove indistinct. Lateral pronotal margins slightly divergent anteriorly, the maximum pronotal width at anterior third. Lateral margins sometimes slightly incurved in middle. Pronotal structure consists laterally of small polygonal cells provided with sharp central grains and of fine transverse wrinkles medially. Scutellum subcordiform, slightly wider than long. Elytra flat, 1.8–2.0 times as long as wide at humeral level, with fine granular sculpture and a silky shine. Apical part of elytral margins distinctly serrate. ♂: mid and hind tibiae serrate on inner margin; anal sternite simply rounded. ♀: mid and hind tibiae simple; anal sternite with a deep apical notch. Length: 7.0–11.0 mm.

Distribution. Not yet found in Denmark or Fennoscandia. The species ranges from Spain to Asia Minor and Caucasus, and from Algeria and South Europe to the Baltic region of the USSR: Estonia, Lithuania, and Latvia.

Biology. The larva develops under bark of branches and thin trunks of *Ulmus laevis* and *U. carpinifollia.* It pupates in the autumn in the superficial layers of the wood. The adult beetle hibernates in the pupal chamber and emerges in May or June. The adults are usually seen on leaves of the host plants, quite exceptionally also on flowers (e. g. of *Crataegus*). The development lasts 2–3 years. Description of larva: Perris (1858), Bilý (1975).

Subgenus *Melanthaxia* Richter, 1944

Melanthaxia Richter, 1944: 119.
Type-species: *Anthaxia godeti* Gory & Laporte de Castelnau, 1841

Black or dark bronze species; body rather wide and subparallel; development in coniferous trees.

Key to species of *Melanthaxia,* adults

| 1 | Head with at least some long distinct pubescence; pronotum without deep punctiform depressions, at most with 4 shallow and indistinct depressions .... 2 |
| - | Head without pubescence; pronotum with 4 deep, roundish punctiform depressions .... 3 |

2 (1) Head with a white pubescence; pronotum laterally with elongate wrinkles in addition to polygonal ground sculpture; mid tibiae of male notched near apex, and with several small spines on inner margin; body slender, with a rather dense white pubescence; 7.0–11.5 mm .... 17. *similis* Saunders
– Head with a black pubescence; pronotum with only the polygonal ground sculpture; mid tibiae of male simple; body robust and subparallel, with extremely fine black pubescence; 5.0–8.1 mm .... *helvetica* Stierlin
3 (1) Body rather flat (Fig. 49); elytra rather long, about 1.8 times as long as wide, with several irregular and very shallow depressions; frons flat; vertex twice as wide as width of an eye; pronotal depressions usually deep; shiny species, usually with a slight metallic tinge; aedeagus as in Fig. 51; 4.0–8.0 mm ....

15. *quadripunctata* (Linné)
– Body more convex (Fig. 48); elytra shorter, about 1.6 times as long as wide, surface regular, only with a shallow basal depression; frons feebly convex; vertex 2.5 times as wide as width of an eye; pronotal depressions usually shallow; matt species, with at most a silky sheen; aedeagus as in Fig. 50; 3.6–7.0 mm .... 16. *godeti* Gory & Castelnau

15. *Anthaxia (Melanthaxia) quadripunctata* (Linné, 1758)
Plate 2:15. Figs. 3–7, 46, 49, 51.

Buprestis quadripunctata Linné, 1758: 410.
Buprestis punctata Ponza, 1805: 81.
Anthaxia angulicollis Küster, 1851: 28.
Anthaxia angulata Küster, 1850: 30.
Anthaxia quadriimpressa Motschulski, 1859: 226.

A black, rather shiny, glabrous and moderately convex species. Head sometimes with a slight bronze sheen, frons almost flat, without pubescence. Vertex 1.6 ($\male$) – 1.8 ($\female$) times wide as width of an eye. Structure of head consists of a dense network of small roundish cells provided with very indistinct central grains. Pronotum slightly convex, 1.8–2.0 times as wide as long. Lateral pronotal margins subparallel, more or less incurved or notched behind middle, rounded only at anterior third. Pronotum has four deep depressions: two small roundish ones on disc, and two large and broader ones at the lateral margins. Pronotal structure consists of small, densely set, polygonal or oval cells without central grains. This structure is somewhat indistinct at the centre. Medial pronotal groove more or less developed. Scutellum triangular, flattened. Elytra slightly convex, 1.6–1.7 times as long as wide at humeral part. Elytral sculpture consists of coarse, irregular and sparse punctures and wrinkles, and of a fine ground microsculpture. Apical part of elytral margins without serrations. ♂: mid and hind tibiae slightly serrate on inner margin. ♀: mid and hind tibiae without serration. Length: 4.0–8.0 mm.

Distribution. Denmark: very rare; only recent records from SJ (Frøslev plantation a. o.); also a few records decades ago from NEZ. Sweden: records from all districts except for G. Sand. and T. Lpm. Norway: in all districts north to Nsi, except for the districts along the western coast. Finland: in all districts except for LkW, Le and Li in the north. USSR: Vib, Kr, Lr. – From western Europe to Siberia, Mongolia, and the Far East, and from the Balkans to N Scandinavia. Not native to Britain; only one very old record.

Biology: The larva develops under the bark of various coniferous trees, especially *Picea excelsa* and *Pinus* spp. There are seven larval instars, and the development lasts one year under the climatic conditions of Central Europe (Bílý, 1975). The development is probably of longer duration in the northern parts of Scandinavia. The larval instars are distinguishable by the width of the epistome:

instar	I	II	III	IV	V	VI	VII
width of epistome in mm	0.31–0.37	0.39–0.46	0.50–0.53	0.58–0.63	0.68–0.76	0.82	0.91–0.94

All larval instars, as well as the pupa and the adult, are able to hibernate under Central European conditions, and the adults occur from late spring through the whole summer. Descriptions of larva: Saalas (1923), Palm (1962), Soldatova (1970), Bílý (1975).

Note. *A quadripunctata* is, with regard to the shape of the body, the most variable species in the genus *Anthaxia*. The shape of pronotum especially is extremely variable. This holds good not only for the form of the lateral margins, but also for the arrangement and form of the pronotal depressions.

16. ***Anthaxia (Melanthaxia) godeti*** Gory & Laporte de Castelnau, 1841
   Figs. 48, 50.

Anthaxia godeti Gory & Laporte de Castelnau, 1841: 31.
Anthaxia submontana Obenberger, 1930: 552.

A small, black, matt, rather convex, glabrous species. Very similar to *quadripunctata*, the two species being a typical example of so-called "sibling species". *A godeti* differs from *quadripunctata* only by the smaller and more convex body (Fig. 48), which has a matt silky sheen, and by the shape of pronotum, which is more convex and rounded laterally. Also by slight differences in the shape of aedeagus (Figs. 50, 51). The species varies in size, but only very slightly in shape of body. Diagnostic characters were also given by Strand (1962).

Distribution. Not in Denmark or Finland; Sweden: Öl., Gtl., and Vstm.; Norway: recorded from several of the southern districts. – The species ranges from Spain to Central Asia, and from S. Europe to S Scandinavia.

Biology. The larva undertakes its development under the bark of various *Pinus* species. On Öland, Sweden, it has once been reared from *Juniperus*. The development lasts one year, the adults occurring from May to August on yellow flowers. The biology is very similar to that of *quadripunctata,* but *godeti* seems to be a more thermophilous species occurring in lowland areas. Description of larva: Soldatova (1970).

17. *Anthaxia (Melanthaxia) similis* Saunders, 1871

Buprestis morio Herbst, 1801: 235 (praeocc. Fabricius, 1792).
Anthaxia similis Saunders, 1871: 54.

A black, flattened and subparallel species, with relatively long pubescence on the entire body. Frontal pubescence long and white, dorsal pubescence shorter and greyish white. Frons flat, vertex 1.3–1.4 times as wide as width of an eye. Pronotum angulate at posterior third, with sculpture consisting of transverse wrinkles and cells in the middle, and of longitudinal cells provided with large central grains laterally. Pronotum slightly grooved medially, 1.6–1.7 times as wide as long, with two large and shallow laterobasal depressions. Elytra flat, shiny, with a coarse granular sculpture, without apical serrations, 1.8–2.0 times as long as wide at base. Some few specimens, of both sexes, possess a green coloration on frons and vertex. ♂: mid tibiae with a deep preapical incision on inner margin; hind tibiae somewhat flattened, incurved on inner margin at posterior third, and enlarged apically; anal sternite regularly rounded. ♀: mid and hind tibiae simple; anal sternite somewhat produced apically. Length: 7.0–11.5 mm.

Distribution. Not in Denmark. Sweden: not recorded from the southern districts such as Sk., Bl., and Hall., but from Sm., Öl., Gtl., Ög., Sdm., Upl., and Hls.; also absent from the northern districts. Norway: Ø, AK, Bø, VE, TEy, TEi, and AAy. Not in Finland or in the Karelian part of the USSR. – Portugal to Ukraine, N Italy and the Balkans to C. Scandinavia. Also present in the Baltic region of the USSR: Estonia, Lithuania and Latvia.

Biology. The larva undertakes its development under the bark, or in the bark, of various *Pinus* species, *Larix decidua,* and *Picea excelsa.* The adult beetle hibernates in the pupal chamber, and occurs in May–June on yellow flowers. Larva undescribed.

Anthaxia (Melanthaxia) helvetica helvetica Stierlin, 1868

Anthaxia helvetica Stierlin, 1868: 345.

A robust, flattened and broad, black-bronze species. Pubescence very short and black, frons only with short, but dense, brownish black pubescence. Frons very slightly convex, vertex very wide, 2.3–2.5 times as wide as width of an eye. Pronotum very slightly angulate at posterior third, very wide, 2.2–2.4 times as wide as long. Pronotal structure consists of irregular rounded or polygonal cells, which are rather indistinct in medial part, without distinct central grains. Pronotum simply convex, with a slight prescutellar depression and feeble laterobasal depressions, or with two or four rounded pronotal depressions (as in *quadripunctata* and *godeti*). Scutellum slightly wider than long. Elytra rather flat, very wide and short, 1.3–1.5 times as long as wide, without apical serrations, and with a slight silky sheen. ♂: mid and hind tibiae with fine serrations on inner margin; body less robust, more slender. ♀: mid and hind tibiae simple; body more robust and wide. Length: 5.0–8.1 mm.

The variability of this species is very slight in regard to coloration, but is conspicuous in regard to body shape, especially that of the pronotum. This may be simply convex, or may have very distinct depressions (see above).

Distribution. Not in Fennoscandia or Denmark. Baltic Region of the USSR: Latvia. Range: Central Europe, Ukraine, North Balkan. Ssp. *apennina* in Italy.

Biology. The larva develops under the bark of *Abies alba, Picea excelsa, Larix decidua,* and various *Pinus* species. The last instar larva, or the pupa, hibernates in a pupal chamber. The adults occur in May–August on yellow flowers. Larva unknown.

SUBFAMILY CHALCOPHORINAE

Radial cell of wing well developed; sensory pores on the serrate antennal segments 5–11 are scattered all over the surface; scutellum extremely small, almost invisible; posterior pronotal margin almost straight; elytra always without rows of punctures but with longitudinal, shiny, raised areas; large species, over 20 mm long.

Larva: labrum lobate laterally; prothorax at most 1.5 times as wide as mesothorax; pronotal and prosternal fields of reddish asperities always large and oval.

Only one tribe.

Tribe Chalcophorini Lacordaire, 1857

Genus *Chalcophora* Solier, 1833

Chalcophora Solier, 1833: 278.
Type-species: *Buprestis mariana* Linné, 1758.

Very large, elongate and flattened species. Entire body usually dark bronze, sometimes with a coppery or greenish sheen, and a pale tomentum. Pronotum and elytra with longitudinal, smooth and shiny keels and raised areas. Metepimera not covered by the basal prolongation of the first abdominal segment. Antennal sensory pores scattered, not concentrated into sensory pits. Basal segment of hind tarsus about twice as long as segment 2.

There are 26 species in the world fauna. Most species occur in the Nearctic Region, 6 are known to occur in the Palaearctic Region. Two species are members of the European fauna. The Nearctic species have frequently been placed in a separate subgenus, *Texania* Casey.

18. *Chalcophora mariana mariana* (Linné, 1758)
Plate 1:7.

Buprestis mariana Linné, 1758: 409.

The largest European buprestid beetle. Body oval and elongate, rather flattened. Dorsal side black or black-bronze, ventral side copper-coloured. Head, pronotum and elytra with wide shiny longitudinal carinae, interstices between them copper-coloured and provided with a fine and dense puncturation. The medial carina of each elytron interrupted by two wide and shallow, copper-coloured depressions. Newly emerged specimens are usually covered with a light white tomentum. Head with short white pubescence, sternal part of ventral side with long white pubescence. Frons with a sharp and deep medial groove, pronotum 1.3 times as wide as long, with wide and shallow latero-basal depressions. Lateral pronotal margins slightly curved, pronotum widest at base. Scutellum very small, oval. Elytra twice as long as wide, their lateral margins sparsely serrate laterally. Each elytron with a small, sharp spine apically. ♂: anal sternite with a wide and deep apical incision. ♀: anal sternite rounded apically. Length: 21.0–32.0 mm.

Distribution. Denmark: not naturally occurring, but a few older finds of introduced specimens in buildings and timber-yards. Sweden: Sk., Sm., Gtl., Sdm., Upl., Vstm., and Hls. Norway: AK, TEy, AAy. Finland: Ab, N, St, Ta, and Sa. USSR: Vib, Kr. – The nominate ssp. is distributed from W Europe to Siberia (Irkutsk-area), and from Italy and the Balkans to C Scandinavia. Ssp. *massiliensis* Villers in the W. Mediterranean area.

Biology. The larva develops in the wood of dead trunks and roots of *Pinus* and *Picea*. In eastern Sweden the main breeding site is railway sleepers. In this area the pupation takes place in July, the adults hatch in late July–early August and leave their pupal chambers. After the hibernation the adult beetles swarm in May–July the following year, appearing on trunks and needles of the host trees. The development lasts 3–6 years. Description of larva: Richter (1952), Palm (1962).

SUBFAMILY CHRYSOBOTHRINAE

Middle-sized species; fore femora with a large spine on anterior surface; radial cell of wing reduced; antennal sclerites separated from frons by a low ridge; antennal segment 3 at least twice as long as segment 4; eyes strongly convergent dorsally; vertex very narrow; each elytron with three longitudinal carinae and three golden green depressions.

54

Larva: labrum not enlarged laterally; prothorax very wide and rounded, twice as wide as mesothorax; meso- and metathorax much wider than abdomen; pronotal and prosternal fields of reddish asperities always large and oval.

Only one tribe.

Tribe Chrysobothrini Laporte de Castelnau & Gory, 1838

Genus *Chrysobothris* Eschscholtz, 1829

Chrysobothris Eschscholtz, 1829: 9.
Type-species: *Buprestis impressa* Fabricius, 1787.

Small, middle-sized or large, flattened species, usually dark bronze, sometimes with metallic elytral spots or shiny, raised areas, exceptionally metallic green or blue, rarely green with dark elytral spots. Eyes very large, vertex always much narrover than diameter of an eye. Third antennal segment at least 2.5 times as long as segment 4. Antennal sensory pores concentrated into pits. Scutellum triangular, extended posteriorly. Fore femora usually with a large tooth. Radial cell of wing reduced. Claws simple. Antennal sclerites separated by a low ridge.

There are about 300 species in the world fauna; most of them are found in the Nearctic Region. 25 species occur in the Palaearctic Region; of these six occur also in Europe.

The genus was divided by Semenov & Richter (1934) into three subgenera: *Chrysobothris* s. str., *Sphaerobothris* Sem. & Richter, and *Abothris* Sem & Richter. All European species belong in *Chrysobothris* s. str.

Key to species of *Chrysobothris*, adults

1 Elytra with strongly elevated smooth carinae; elytral interstices with dense and irregular wrinkled granular sculpture;
elytra usually with a narrow coppery or purple lateral mar-
gin; 10.0–16.6 mm .... 20. *c. chrysostigma* (Linné)
Elytra with feeble carinae which are indistinct in the basal
part; elytral interstices finely and densely punctured; elytral
margin of the same colour as the disc; 9.0–15.0 mm .... 19. *a. affinis* (Fabricius)

Key to species of *Chrysobothris*, larvae

1 Pronotal sclerotized asperities with a low rounded tooth (Fig.
52); mesothoracic spiracle 2.4 times as wide as long; submaxillary sclerite with two bristles and two sensory areas
(Fig. 54); host plants: almost all deciduous trees .... 19. *affinis* (Fabricius)
– Pronotal sclerotized asperities with a small sharp tooth (Fig.
53); mesothoracic spiracle 2.7 times as wide as long; submaxil-

55

lary sclerite with two sensory bristles only (Fig. 55); host
plants: *Abies, Picea* .... 20. *chrysostigma* (Linné)

Figs. 52, 53. Pronotal asperities of larva of 52: *Chrysobothris affinis* (F.) and 53: *C. chrysostigma* (L.).
Figs. 54, 55. Submaxillary sclerite of larva of 54: *Chrysobothris affinis* (F.) and 55: *C. chrysostigma* (L.).

19. *Chrysobothris (Chrysobothris) affinis affinis* (Fabricius, 1794)
Figs. 52, 54.

Cucujus chrysostigma Fourcroy, 1785: 32 (nec Linné).
Buprestis affinis Fabricius, 1794: 450.
Buprestis congener Paykull, 1799: 222.
Chrysobothris assimulans Schreibers, 1843: 61.

A flat and oval species. Dorsal side bronze, with three pairs of copper-coloured elytral depressions; ventral side copper-coloured. Frons green (♂) or bronze (♀), flat with a transverse elevation or feeble carina on upper part. Vertex convex, 0.4 times as wide as width of an eye. Eyes large, slightly projecting beyond the outline of the head. Pronotum twice as wide as long, with almost parallel lateral margins. Pronotal pubescence somewhat more sparse than pubescence of head. Structure of pronotum consists of coarse copper-coloured puncturation and fine transverse wrinkles. Scutellum small, green and triangular. Elytra rather flat, oval, 1.7–1.8 times as long as wide, with fine simple puncturation. Each elytron with three very feeble longitudinal carinae, and three shallow copper-coloured depressions. The puncturation of these depressions is finer and denser than that of the elytra. Apical elytral margins slightly serrate. ♂: mid tibiae bent; anal sternite with a deep and wide triangular incision. ♀: mid tibiae only very slightly bent; anal sternite slightly bisinuate apically. Length: 9.0–15.0 mm.

Distribution. Denmark: LFM, SZ, NEZ, very rare. Sweden: from Sk. north to Upl. and Vstm. Norway: only in some districts along the southern coast: VE, TEy, AAy, and VAy. Not in Finland, nor in the Karelian part of the USSR. – Throughout Europe, but not in Great Britain, eastwards to Siberia. Other ssp. in NW Africa and Asia.

Biology. An extremely polyphagous species, developing in practically every sort of deciduous trees, in our area especially *Quercus, Fagus,* and *Betula.* The larva tunnels under the bark, and pupates in the wood in early spring. The adults appear in May–July. The development lasts 2–3 years. Description of larva: Schiødte (1870), Perris (1877), Richter (1952), Palm (1962), Bílý (1975).

20. *Chrysobothris (Chrysobothris) chrysostigma chrysostigma* (Linné, 1758).
Plate 1:3: Figs. 53, 55.

Buprestis chrysostigma Linné, 1758: 409.

An oval, rather flattened species. Dorsal side black or dark bronze, with copper-coloured elytral margins and three pairs of elytral spots. Frons and anterior pronotal margin green with a metallic sheen, ventral side golden green with a blue tinge, lateral parts of abdomen purple. Head, pronotum and ventral side with rather long white pubescence, elytra glabrous. Head relatively small, frons flat, vertex very convex, 0.5 times as wide as width of an eye. Pronotum 1.7–1.8 times as wide as long, with coarse golden green puncturation and transverse wrinkles in the medial part. Medial pronotal groove shallow and fine. Scutellum small, black, triangular. Elytra 1.8–1.9 times as long as wide, oval. Each elytron with 4 shiny carinae and 3 deep, golden green or coppery depressions. Basic sculpture of elytra coarse, but shiny. Lateral elytral margins slightly serrate at apex. ♂: fore and mid tibiae bent; anal sternite with a wide and deep incision; ♀: fore and mid tibiae only slightly bent; anal sternite only slightly incurved apically. Length: 10.0–16.5 mm.

Distribution. Denmark: NEZ, Fredensborg, 1 specimen 1793, probably representing a natural, now extinct, population; also introduced, but rarely. Sweden: records from nearly all districts from Sk. to T. Lpm. Norway: only records from the SE parts of the country, not known from the western parts. Finland: in all districts, except for Al, Le, and Li; USSR: Vib and Kr. – The nominate ssp. occurs from West Europe to Lake Baikal in Siberia, from Italy to N Fennoscandia. Other ssp. occur in the Far East, Japan, and Greece.

Biology. The larva undertakes its development under the bark of dying or dead trunks and branches of *Abies alba, Picea excelsa* and *Pinus sylvestris.* The pupal chamber is formed in the wood; the development lasts 2–3 years. The adults appear on the host plants in June–August. Description of larva: Richter (1952), Palm (1962), Bílý (1975).

SUBFAMILY AGRILINAE

Elongate, subcylindrical species; frons convex or grooved medially; both sections of mesosternum well developed; hind coxae enlarged exteriorly; claws with a large tooth; elytra more than 2.5 times as long as wide.

Larva: prothorax moderately enlarged (Fig. 8); pronotum and prosternum with a reddish sclerotized groove (Agrilini), or without such groove (Aphanisticini); anal segment well developed, with two sclerotized spines (Agrilini), or small and without spines (Aphanisticini).

Key to tribes and genera of Agrilinae, adults

1   Scutellum with a transverse carina (Fig. 57), except in *Agrilus subauratus* (Fig. 58); basal segment of hind tarsus long and slender, much longer than segments 2 & 3 together (Fig. 59); pronotum with two lateral carinae (Figs. 61–63), and with transverse wrinkles; prosternum and femora without grooves for the reception of the antennae and tibiae; frons flat or convex; tarsi with two claws (Tribe Agrilini) .... *Agrilus* Curtis (p. 59)

–   Scutellum without a transverse carina; basal segment of hind tarsus only slightly longer than segment 2 (Fig. 60); pronotum with a simple lateral carina, without transverse wrinkles, only with a simple, fine and sparse, puncturation; prosternum and femora with grooves for the reception of the antennae and the tibiae; frons with a shallow, but wide, medial grove (Fig. 56); tarsi with only one claw (Tribe Aphanisticini) .... *Aphanisticus* Latreille (p. 84)

Key to genera of Agrilinae, larvae

1   Prothorax with longitudinal grooves, which are more-or-less well sclerotized (Figs. 8, 86–88); anal segment with two sclerotized spines (Figs. 8, 89, 90) .... *Agrilus* Curtis (p. 59)

–   Prothorax without grooves; anal segment simple ...... *Aphanisticus* Latreille (p. 84)

Tribe Agrilini Laporte de Castelnau et Gory, 1839

Middle- to small-sized, elongate species, which are acuminate posteriorly; scutellum with a transverse keel (Fig. 57), except in *A. subauratus* (Fig. 58); basal segment of hind tarsus long and slender, much longer than segment 2 & 3 together (Fig. 59); pronotum with two lateral carinae (Figs. 61–63), and with transverse wrinkles; prosternum and femora without grooves for the reception of the antennae and tibiae; frons flat or vaulted; tarsi with two claws.

Larva: prothorax with longitudinal sclerotized grooves (Fig. 8); anal segment with two sclerotized spines (Figs. 8, 89, 90).

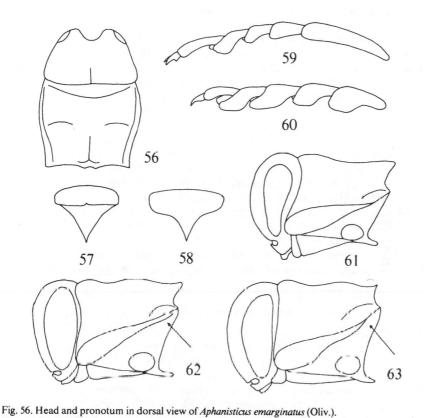

Fig. 56. Head and pronotum in dorsal view of *Aphanisticus emarginatus* (Oliv.).
Figs. 57, 58. Scutellum of 57: *Agrilus sulcicollis* Lac. and 58: *A. subauratus* Gebl.
Figs. 59, 60. Hind tarsus of 59: *Agrilus sulcicollis* Lac. and 60: *Aphanisticus pusillus* (Oliv.).
Figs. 61–63. Head and prothorax in lateral view of 61: *Agrilus sulcicollis* Lac.; 62: *A. viridis* (L.);
63: *A. mendax* Mannh.

Genus *Agrilus* Curtis, 1825

Agrilus Curtis, 1825: 67.
  Type-species: *Buprestis viridis* Linné, 1758.

Teres Harris, 1829: 2.
Anambus Thomson, 1864: 38.
Diplolophotus Abeille de Perrin, 1897: 2.
Uragrilus Semenov, 1935: 276.
Epinagrilus Stepanov, 1953: 114.

Small to middle-sized species of elongate subcylindrical or acuminate form (Figs. 1, 2). Coloration very variable, from black to metallic green or blue; elytra and pronotum sometimes with white, yellow, or orange, tomentose spots. Both sections of mesosternum reduced, almost indistinct. Prosternum always with an anterior prosternal lobe. Exterior part of metacoxae enlarged. Claws with a large tooth. Scutellum (Figs. 57, 58) transverse, with a posterior spine and a transverse keel (except in *A. subauratus*). Basal segment of hind tarsus (Fig. 59) at least as long as segments 2–4 together. Pronotum with two lateral carinae(Figs. 61–63), i.e. with an upper and lower carina. Posterior pronotal angles usually with a prehumeral carina.

Larva: prothorax with longitudinal medial groove, which is more or less sclerotized (Figs. 8, 86–88); anal segment with two sclerotized spines (Figs. 8, 89, 90).

An extremely large and taxonomically difficult genus distributed all over the world. About 2,500 species have been described. The palaearctic fauna comprises about 400 species, 80 of which are known to occur in Europe.

All attempts to divide this genus into subgenera have failed so far. The palaearctic fauna of this genus was divided into species-groups by Schaefer (1949) and Bilý (1977). The oriental species were grouped by Baudon (1968).

In the *A. viridis*-group the observed variability in respect to both morphology and biology suggests that this complex is still undergoing evolutionary differentiation.

Key to species of *Agrilus*, adults

1   Elytra with two or six white tomentose spots; also pleurites
of abdominal segments with white tomentose spots 2
–   Elytra and abdominal pleurites without white tomentose
spots 4

2 (1)   Elytra rounded apically (Fig. 64), with two white spots at
posterior third; last abdominal tergite in both sexes
rounded, without a carina; 8.3–13.0 mm .... 26. *biguttatus* (Fabricius)
–   Elytra pointed apically (Figs. 65, 66), with six white spots;
last abdominal tergite in both sexes with a medial carina
extending beyond posterior margin in the form of a blunt
spine (Fig. 67) 3

3 (2)   Lateral and sutural margins of elytra convergent at the
apex which possesses a small spine (Fig. 65); prehumeral
pronotal carinae sharp, short and arched; 6.6–11.0 mm .... 22. *a. ater* (Linné)
–   Lateral margin of elytra straight or slightly excurved at
apex, the sutural margins strongly divergent at apex (Fig.
66); prehumeral pronotal keels indistinct; 8.5–12.0 mm

                                             29. *guerini* Lacordaire

4 (1)   Posterior margin of last abdominal sternite more-or-less
incurved, or at least sinuous (Fig. 68) 5
–   Last abdominal sternite rounded apically (Fig. 69) 11

5 (4) Scutellum without or with only a poorly defined transverse keel (Fig. 58); elytra golden green and pronotum blue-green or whole body blue; large species, 6.5–10.0 mm .... 36. *subauratus* Gebler
  – Scutellum with a well defined transverse keel (Fig. 57); smaller species 6
6 (5) Frons and vertex deeply grooved medially; prehumeral pronotal keels absent or only slightly indicated; a blue or blue-green species with a silky sheen; 4.5–7.5 mm ..... 28. *cyanescens* Ratzeburg
  – Frons and vertex with only a slight longitudinal depression; prehumeral pronotal keels well developed 7
7 (6) Metasternum depressed between mid coxae; prosternal process depressed; pronotum very convex, its anterior and posterior margins of the same width; prehumeral pronotal keels sharp and straight; a small, brownish green species with a brassy tinge; 3.5–5.5 mm .... 27. *c. convexicollis* Redtenbacher
  – Metasternum not grooved between hind coxae; prosternal process flat; pronotum enlarged anteriorly, with a medial groove, or with several depressions 8

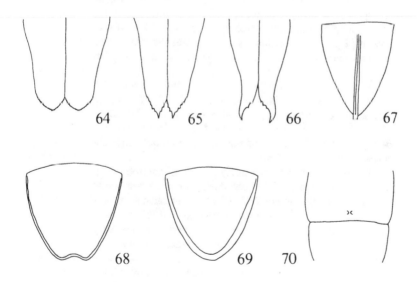

Figs. 64–66. Apex of elytra of 64: *Agrilus biguttatus* (F.); 65: *A. ater* (L.); 66: *A. guerini* Lac.
Fig. 67. Last tergite of *Agrilus ater* (L.).
Figs. 68, 69. Anal sternite of 68: *Agrilus subauratus* Gebl. and 69: *A. viridis* (L.).
Fig. 70. Sternite II of male of *Agrilus olivicolor* Kiesw.

8 (7) Entire surface of elytra clothed with fine white pubescence; slender and elongate species; prehumeral pronotal carinae straight; second abdominal sternite of male with two small bumps near the mid line (Fig. 70); prosternal process with white pubescence; 4.5–5.5 mm .... 33. *olivicolor* Kiesenwetter
–   Elytra glabrous 9
9 (8) Prosternal process widening behind hind coxae (Fig. 71); elytra sometimes with several small white scales along suture; male antennae extremely enlarged from segment 4 onwards; a green or brownish green species; aedeagus as Fig. 73; 4.3–6.2 mm .... 31. *laticornis* (Illiger)
–   Prosternal process parallel–sided, not widening behind hind coxae (Fig. 72); elytra completely glabrous; male antennae not enlarged 10
10 (9) Vertex narrow, about 1/3 the width of the anterior pronotal margin; eyes weakly convex; prehumeral pronotal carinae short and arched; elytra more robust; a large, green, golden-green, bronze, blue or violet species; aedeagus as in Fig. 75; 6.0–8.5 mm .... 37. *sulcicollis* Lacordaire
–   Vertex wider, about ½ the width of the anterior pronotal margin; eyes strongly convex; prehumeral pronotal carina sharp, long and straight, reaching almost to middle of pronotum; elytra more slender; a smaller, green or blue–green species (rarely blue or bronze); aedeagus as in Fig 74; 3.7–6.5 mm .... 21. *angustulus* (Illiger)
11 (4) Elytra entirely clothed with fine white pubescence; prehumeral pronotal carinae not or only slightly developed; pronotum almost cylindrical, with deep and wide medial depressions in posterior half, but lateral depressions absent; vertex with dense and regular puncturation; reddish-bronze species; 3.5–6.5 mm .... *hyperici* (Creutzer)
–   Elytra glabrous, or with extremely fine, indistinct pubescence 12
12 (11) Anterior prosternal lobe protruding, with a narrow medial notch (Fig. 83); lower and upper lateral carinae of pronotum confluent in posterior third (Fig. 63); elytra completely glabrous; each elytron with a very fine not very obvious medial, longitudinal carina; a large, golden-bronze species; 10.0–12.5 mm .... 32. *mendax* Mannerheim
–   Anterior prosternal lobe not so protruding, only with a very shallow medial emargination (Fig 84); coloration different; both lateral carinae of pronotum free, not confluent in posterior part (Fig. 62) 13
13 (12) Prehumeral pronotal carinae indistinct; vertex grooved

medially; eyes small; vertex wide, 2/3 the width of the anterior pronotal margin (Fig. 76); pronotum with a deep and wide medial depression which is nearly interrupted in middle; basal segment of hind tarsi as long as segments 2 & 3 together; a robust, brown or black-brown species, with a bronze lustre; 5.0–7.5 mm .... 30. *integerrimus* (Ratzeburg)

– Prehumeral pronotal carinae well developed 14

14 (13) Pronotum with deep and wide depressions at lateral margins; vertex with fine spiral or simple puncturation 15

– Pronotum without deep lateral depressions; vertex with fine longitudinal wrinkles (Fig. 77) 17

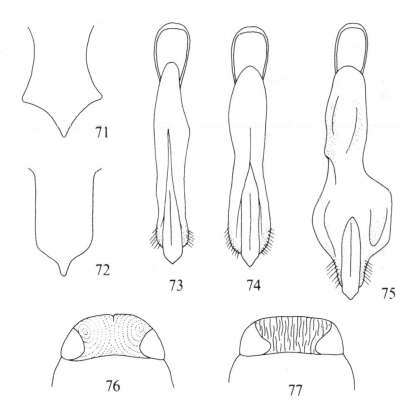

Figs. 71, 72. Prosternal process of adult of 71: *Agrilus laticornis* (Ill.) and 72: *A. sulcicollis* Lac.
Figs. 73–75. Aedeagus of 73: *Agrilus laticornis* (Ill.); 74: *A. angustulus* (Ill.); 75: *A. sulcicollis* Lac.
Figs. 76, 77. Head of 76: *Agrilus integerrimus* (Ratz.) and 77: *A. viridis* (L.).

63

15 (14) Pronotum with coarse transverse wrinkles, which are broken up here and there by coarse punctures; vertex very convex; head and pronotum golden yellow or reddish bronze, elytra green or blue-green; 4.0–6.5 mm ...... 34. *p. pratensis* (Ratzeburg)
- Pronotum with very fine sculpture of closely spaced, fine, transverse wrinkles; vertex slightly convex or flat; unicolourous species (blue, dark green, or black) 16

16 (15) Head and pronotum with fine microsculpture and a matt silky sheen; prehumeral pronotal carinae long, about 1/3 of pronotal length; longitudinal medial depression of pronotum shallow, but distinct; a longer, dark olive-green or black species; 4.0–6.0 mm .... 25. *b. betuleti* (Ratzeburg)
- Head and pronotum without microsculpture, shiny; prehumeral pronotal carinae sharp but short, at most about ¼ of pronotal length; longitudinal medial depression of pronotum only slightly developed at posterior part; a blue species; 4.0–5.5 mm .... 35. *p. pseudocyaneus* Kiesenwetter

17 (14) Long and slender species with elytra acuminate apically; apex of elytra distinctly serrate (Fig. 80); vertex almost flat; lateral pronotal margins always slightly incurved before posterior corners, which are right-angled or acute; dark green, blue, blue-green, violet, or bicolourous species 18
- Smaller and more robust species with very feebly acuminate elytra; elytral apex indistinctly serrate (Fig. 81); vertex very convex; lateral pronotal margins straight or very slightly incurved before posterior corners, these being obtuse or at most right-angled; black, dark green, green-bronze, green, or slightly bicolourous species 19

18 (17) Large and slender species with apex of elytra very elongated; mid femora serrate posteriorly (Fig. 79); tarsi, especially hind tarsi, gradually enlarged from base to segment 4; third tarsal segment about as wide as segment 4 (Fig. 79); a light green, blue-green, or, rarely, golden green species; head and pronotum in rare cases with a bronze tinge; 7.0–10.5 mm .... 38. *suvorovi populneus* Schaefer
- Smaller, less elongate species with a less acuminate elytral apex; mid femora with a smooth posterior margin (Fig. 78); tarsi slender, segments 1 and 2 of equal width, segment 3 somewhat wider, and segment 4 enlarged (Fig. 78); extremely variable species both in regard to size and coloration: from black through all grades of green and blue to violet; 4.5–11.5 mm .... 39. *v. viridis* (Linné)

19 (17) Lateral pronotal margins slightly incurved before posteri-

or corners which are almost right-angled; frons and vertex very convex; elytral apex very slightly, but distinctly, serrate; pronotum 1.7–1.8 times as wide as long; a black or black bronze species; 5.4–7.0 mm .... 24. *aurichalceus paludicola* Krogerus
–   Lateral pronotal margins straight, not incurved before posterior corners which are obtuse; frons and vertex less convex; elytral apex indistinctly serrate, almost smooth; pronotum almost twice as wide as long; a green, brownish green, green-bronze, or slightly bicolourous, species; head and pronotum golden green and elytra green; 4.5–7.0 mm

23. *a. aurichalceus* Redtenbacher

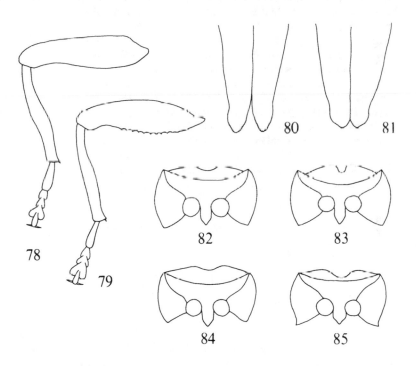

Figs. 78, 79. Left mid leg of 78: *Agrilus viridis* (L.) and 79: *A. suvorovi populneus* Schaef.
Figs. 80, 81. Apex of elytra of 80: *Agrilus suvorovi populneus* Schaef. and 81: *A. aurichalceus* Redt.
Figs. 82–85. Prosternum of 82: *Agrilus sulcicollis* Lac.; 83: *A. mendax* Mannh.; 84: *A. viridis* (L.); 85: *A. cyanescens* Ratz.

65

Key to species of *Agrilus*, larvae

1   Pronotal groove bifurcate posteriorly (Figs. 86, 87) 2
–   Pronotal groove simple, not bifurcate posteriorly 5
2 (1) Pronotal groove reddish brown anteriorly, without pigmentation posteriorly; pubescence of anal segment dense and
     long, hairs up to half as long as anal spines (Fig. 89) 3
–   Entire pronotal groove reddish brown or yellowish brown;
     pubescence of anal sternite short and sparse, hairs less than
     half as long as anal spines (Fig. 90) 4
3 (2) Pronotal groove enlarged anteriorly (Fig. 86); anal spines
     wide (Fig. 91); host plant: *Betula* .... 25. *betuleti* (Ratzeburg)
–   Pronotal groove not enlarged anteriorly (Fig. 87); anal
     spines narrow (Fig. 92); host plants: *Salix caprea, Populus* ... 36. *subauratus* Gebler
4 (2) Pigmentation of pronotal groove dark posteriorly and paler
     anteriorly, but whole groove reddish brown; anal spines
     narrow (Fig. 97); host plant: *Quercus* .... 26. *biguttatus* (Fabricius)
–   Pronotal groove reddish brown posteriorly and yellow anteriorly; anal spines wide (Fig. 98); host plant: *Salix* .... 22. *ater* (Linné)
5 (1) Pronotal groove enlarged posteriorly; prosternal groove
     enlarged both posteriorly and anteriorly (Fig. 88); anal
     spines narrow (Fig. 94); host plant: *Sorbus aucuparia* ..... 32. *mendax* Mannerheim
–   Pronotal groove not enlarged posteriorly; if enlarged, then
     prosternal groove not enlarged at both ends 6
6 (5) Pronotal and prosternal grooves wide, ratio of width to length 1:12 7
–   Pronotal groove narrow, ratio of width to length from 1:15
     to 1:20 8

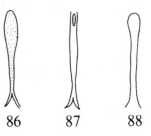
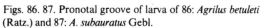

86   87   88

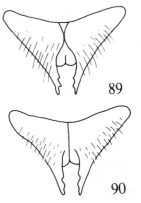

89

90

Figs. 86. 87. Pronotal groove of larva of 86: *Agrilus betuleti*
(Ratz.) and 87: *A. subauratus* Gebl.
Fig. 88. Prosternal groove of larva of *Agrilus mendax*
Mannh.
Figs. 89, 90. Anal segment of larva of 89: *Agrilus betuleti*
(Ratz.) and 90: *A. biguttatus* (F.).

7 (6) Pronotal groove narrowed in posterior fourth; prosternal groove enlarged anteriorly; anal spines narrow (Fig. 96); host plant: *Lonicera nigra* .... 28. *cyanescens* (Ratzeburg)

–   Pronotal groove somewhat narrowed in anterior fourth; prosternal groove almost parallel-sided, only very slightly narrowed both anteriorly and posteriorly; anal spines wide

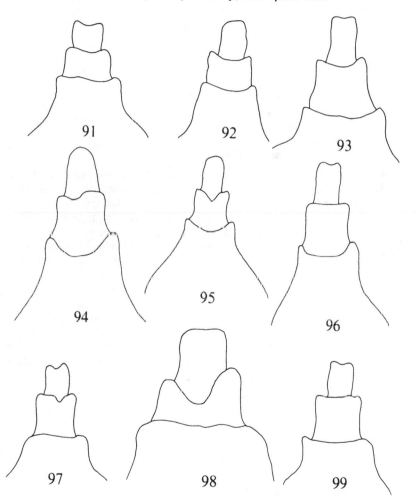

Figs. 91–99. Anal spines of *Agrilus*-larvae, lateral view. – 91: *betuleti* (Ratz.); 92: *subauratus* (Gebl.); 93: *viridis* (L.); 94: *mendax* Mannh.; 95: *aurichalceus* Redt.; 96: *cyanescens* Ratz.; 97: *biguttatus* (F.); 98: *ater* (L.); 99: *integerrimus* (Ratz.) (After Alexeev, 1961).

(Fig. 99); host plant: *Daphne mezereum* .... 30. *integerrimus* (Ratzeburg)

8 (6) Prosternal groove strongly sclerotized and pigmented posteriorly; anal spines narrow (Fig. 93); host plants: *Quercus,
Fagus, Salix, Betula, Populus, Carpinus* etc. .... 39. *viridis* (Linné)

– Entire prosternal groove regularly sclerotized and pigmen-
ted; anal spines narrow (Fig. 95); host plant: *Salix* 23. *aurichalceus* Redtenbacher

21. *Agrilus angustulus* (Illiger, 1803)
Fig. 74.

Buprestis angustulus Illiger, 1803: 240.
Buprestis olivaceus Gyllenhal, 1808: 454 (nec Ratzeburg).
Agrilus viridis Stephens, 1830: 174.
Agrilus laetefrons Mannerheim, 1837: 115.
Agrilus rugicollis Ratzeburg, 1837: 55.
Agrilus pavidus Gory & Laporte de Castelnau, 1841: 53.

A small slender species. Dorsal side green or olive-green, pronotum in rare cases golden green. Ventral side green-bronze, or black with a bronze lustre. Pubescence of entire body, except for frontal part, extremely short and indistinct. Frons and vertex slightly convex, with a fine medial groove. Eyes rather large, slightly projecting beyond outline of head in male, not projecting at all in female. Pronotum 1.4–1.5 times as wide as long, almost regularly widening anteriorly, with straight lateral margins. Anterior pronotal margin slightly lobate in middle, medial pronotal groove developed only in posterior half of pronotum. Prehumeral pronotal carinae sharp, straight and long, reaching at least to middle of pronotum. Elytra slightly convex, 3 times as long as wide at humeral level, obtusely rounded, very finely serrate apically. Anal sternite slightly in-curved apically, anterior prosternal lobe notched. Prosternal process flat and parallel-sided, not widening behind fore coxae, mesosternum not grooved. ♂: abdominal ster-nite II with two small warts; frontal and prosternal pubescence denser. ♀: abdominal sternite II smooth, without warts; frontal and prosternal pubescence sparser. Length: 3.7–6.5 mm.

Distribution. Denmark: widespread but more local in the west than in the south-east; known from all districts except for NWJ and B. Sweden: in all districts from Sk. north to Vstm. and Upl. Norway: only in the districts in the southeastern parts: Ø, AK, Bø, VE, TEy, AAy, VAy. Finland: only Ab and Ka. USSR: ? Vib. – From Great Britain through entire Europe to W Siberia, Caucasus and Asia Minor, and from N Africa to C Fennoscandia.

Biology. The larva develops under the bark of thin branches and twigs of various *Quercus* species, *Fagus sylvatica*, *Carpinus betulus*, *Corylus avellana* and *Castanea sativa*. One of the most common European buprestids. The development lasts one year, adults appearing in May-July. Description of larva: Perris (1877), Alexeev (1981).

68

22. **Agrilus ater ater** (Linné, 1767)
   Figs. 65, 67, 98.

Buprestis atra Linné, 1767: 663.
Buprestis sexguttatus Brahm, 1790: 141.
Buprestis biguttatus Rossi, 1790: 191 (nec Fabricius).

A large cylindrical species with eight white tomentose spots, six on elytra and two on pronotum. Dorsal side blue-green or black-blue, ventral side black-blue, sometimes with a violet lustre. Frons almost flat, vertex grooved medially, whole head with short and sparse white pubescence. Pronotum 1.5 times af wide as long, lateral depressions with a snow-white tomentum. Prehumeral pronotal carinae short, bent but sharp, and approaching lateral pronotal margins anteriorly. Transverse scutellar carina well developed. Elytra 3 times as long as wide, subcylindrical and slightly acuminate in posterior third. Each elytron slightly serrate apically, with a sharp medial spine. Prosternal process with parallel margins, anal sternite rounded. Laterotergites with large snow-white tomentose spots. Abdominal sternites with smaller, white or cream-white, tomentose spots. ♂: eyes large, more convex; sternites II and III with a shallow medial groove; frontal and prosternal pubescence dense and long. ♀: eyes smaller, almost flat; sternites II and III without a medial groove; frontal and prosternal pubescence sparser and shorter. Length: 6.5–11.0 mm.

Distribution. Finland: Ab, N. The nominate ssp. distributed in South and Central Europe, in the Baltic region of the USSR: Estonia, Lithuania, Latvia, also in Ukraine, Caucasus and Asia Minor.

Biology. The larva undertakes its development under the bark of trunks or thick branches of *Populus nigra*, *P. alba*, *P. tremula*, *Salix alba* and *S. cuprea*. The larval tunnels are undulating and irregularly situated in the trunk. Pupation takes place in May-June in the superficial part of the wood, or in thick bark. The adults can be found in June-July on leaves of the host plants. Description of larva: Kangas (1947), Alexeev (1960, 1961, 1981), Plochich (1969).

23. **Agrilus aurichalceus aurichalceus** Redtenbacher, 1849.
   Figs. 81, 95.

Agrilus aurichalceus Redtenbacher, 1849: 286.
Agrilus proximus Rey, 1891: 19.
Agrilus epistomalis Abeille de Perrin, 1897: 4.
Agrilus rubicola Abeille de Perrin, 1897: 5.
Agrilus chrysoderes Bétis, 1914: 457 (nec Abeille).
Agrilus politus Weiss, 1914: 438.

A small, somewhat robust species. Entire body green, olive-green, golden green, or blue-green, ventral side usually somewhat darker. Dorsal side practically glabrous, ventral side with extremely fine and sparse white pubescence; frons and prosternum, especially of the male, with short white pubescence. Head large, frons and vertex slightly convex, vertex with a shallow and indistinct medial groove, and with longitudinal lines composed of small punctures. Pronotum very wide, almost twice as wide as long, with regularly rounded lateral margins which are not incurved before posterior angles. Maximum pronotal width at anterior third. Medial pronotal groove feeble, divided into a larger prescutellar and a smaller anterior depression. Prehumeral pronotal carinae short and little elevated, but sharp. Elytra rather robust, 3.2-3.5 times as long as wide at base, somewhat widened at posterior third. Elytral apex slender, indistinctly serrate, almost smooth, each elytron separately rounded. Anal sternite rounded apically, anterior prosternal lobe arcuate. ♂: eyes large, somewhat projecting beyond outline of head; medial antennal segments with a sharp outer angle; frontal and prosternal pubescence rather long and dense. ♀: eyes smaller, not projecting beyond outline of head; medial antennal segments with an obtuse outer angle; frontal and prosternal pubescence shorter, almost indistinct. Length: 4.5-7.5 mm.

Distribution. Sweden: Bl.; Finland: Sa (Imatra). – Baltic region of the USSR: Estonia, Lithuania, Latvia. The nominate ssp. ranges from Portugal to Siberia and Caucasus and from North Africa to South Scandinavia; introduced to the U. S. A. See also no. 24.

Biology. The larva bores spiral tunnels under the bark of various species of *Rosa* and *Rubus*. The development lasts one year; the pupal chamber is made in the wood. The pupation takes place in May-June, adults appearing in May-September on leaves and twigs of the host plants. Description of larva: Perris (1877), Alexeev (1960, 1961, 1981).

Note. This species shows the same type of variability (except in coloration) as in *viridis* (L.). Many taxonomic problems have to be solved especially in regard to the southern and eastern subspecies.

24. *Agrilus aurichalceus paludicola* Krogerus, 1922, syn. nov.

Agrilus viridis paludicola Krogerus, 1922: 110.

A small, rather robust species. Entire body black, sometimes with a brassy tinge, especially of the ventral side. Dorsal and ventral surface of body with indistinct pubescence; frontal and prosternal pubescence white and short, very sparse. Head rather large; frons slightly, and vertex very, convex. Vertex with longitudinal lines composed of fine punctures. Eyes large, not projecting beyond outline of head. Pronotum wide, 1.7-1.8 times as wide as long, with regularly curved lateral margins. Maximum pronotal width at anterior quarter; lateral pronotal carinae slightly incurved before posterior angles. Median pronotal groove divided into a large, wide prescutellar depression and a smaller, but distinct, rounded anterior depression. Prehumeral pronotal carinae sharp, slightly bent, reaching the anterior third of pronotum. Elytra somewhat flattened, 3.0 times as long as wide at base, slightly widened at posterior

70

third. Apical part of elytra slender, distinctly serrate; each elytron separately and narrowly rounded. Anal sternite rounded apically, anterior prosternal lobe arcuate. ♂: eyes larger; frons with a green sheen; medial antennal segments with sharp outer angle; frontal and prosternal pubescence rather well developed. ♀: eyes smaller; frons with a bronze sheen; medial antennal segments with obtuse outer angle; frontal and prosternal pubescence less developed. Length: 5.4-7.0 mm.

Distribution. Sweden: only in the north, Hrj., Vb., Nb., Lu. Lpm., and T. Lpm. Norway: only On. Finland: recorded from Ab, N, St, Sa, Tb, Sb, Kb, Ok, ObN, and LkW. USSR: Vib, Kr, and Lr. – Also in the Estonian SSR.

Biology. The larva undertakes its development under the bark of *Betula nana* (Lundberg, 1960). The development probably lasts one year; pupation in June. Adults in June-August. They have been taken on leaves of *Betula*, and have also been captured in window-traps operating in swamps. Larva undescribed.

Note. *Agrilus aurichalceus paludicola* is sympatric with *A. aurichalceus aurichalceus* in the southeastern part of Finland and in Estonia. The nominal subspecies is very rare in our area, and is replaced in most districts by *A. aurichalceus paludicola*. Some authors (e. g. Obenberger, 1927, 1936 and Alexeev, 1960) consider *paludicola* to be a distinct species. After having studied the type material and extensive material from various Scandinavian institutions, I conclude that it is a northern subspecies of *A. aurichalceus*. Although originally described as a subspecies of *A. viridis*, it belongs to *A. aurichalceus* according to the form of the pronotum, the aedeagus and the elytral apex. Also the extensively sympatric distribution with *A. viridis viridis* is one of the reasons why it cannot be accepted as a subspecies of *viridis*.

25. *Agrilus betuleti betuleti* (Ratzeburg, 1837).
Figs. 86, 89, 91.

Buprestis betuleti Ratzeburg, 1837: 57.
Agrilus foveicollis Marseul, 1869: 122, syn. nov.

A small, rather robust species with a silky sheen. Entirely black, elytra sometimes with a slight bronze or green tinge. Pubescence of entire body quite indistinct. Head small and convex; eyes small and not projecting beyond outline of head. Frons and vertex grooved medially; vertex with fine concentric grooves consisting of fine and small punctures. Pronotum very wide, 1.8-2.0 times as wide as long. Lateral pronotal margins lobate near middle and incurved before posterior angles. Lateral pronotal depressions large and wide. The fine medial pronotal groove divided into elongate depressions anteriorly and posteriorly. Prehumeral pronotal carinae short, but sharp and elevated. Entire pronotum with fine microsculpture, matt. Elytra rather robust, slightly flattened, 3 times as long as wide at base. Apical part of elytra moderately acuminate, without apical serrations; each elytron simply rounded. ♂: frons greenish; tibiae slightly incurved on inner margin. ♀: frons black or black-bronze; tibiae not incurved. Length: 4.0–6.0 mm.

Distribution. Not in Denmark. Sweden: in most districts from Sk. in the south to Hls. in the north. Norway: only Bø. Finland: Ab, N, Ka, St, Sa, and Kb. USSR: Vib and Kr. – The nominate ssp. is found from the Baltic region of the USSR and Central Europe eastwards to Siberia, south to North Italy.

Biology. The larva develops under the bark of branches and twigs of various *Betula* species. It pupates in the wood in May-June, the adult appearing in June-August on leaves of the host plants. The development lasts one year. Description of larva: Alexeev (1960, 1961, 1981).

Note. I have studied the holotype of *Agrilus foveicollis* Marseul, 1869, and found it to be conspecific with *Agrilus betuleti betuleti* (Ratzeburg, 1837), thus a junior synonym. More than 300 specimens were studied from the whole distributional area.

26. *Agrilus biguttatus* (Fabricius, 1777).
   Figs. 8, 64, 90, 97.

Buprestis biguttatus Fabricius, 1777: 137 (nec Rossi).
Buprestis pannonicus Piller & Mitterpacher, 1783: 92.
Cucujus octoguttatus Fourcroy, 1785: 33.
Buprestis subfasciata Ménétries, 1832: 153.
Anambus caeruleoviolaceus Thomson, 1864: 39.

A large, robust and somewhat cylindrical species. Dorsal side golden green, green, blue-green, blue, or violet; elytra with two small white tomentose spots at posterior fourth. Ventral side blue-green, blue, or very rarely violet. Lateral parts of sternites, laterotergites and, rarely, anterior pronotal angles, with areas of white tomentum. Both dorsal and ventral pubescence very short and sparse, only head with distinct white pubescence. Frons with a wide and flat, shallow depression; vertex slightly grooved medially. Pronotum 1.6–1.7 times as wide as long, with two shallow depressions medially: a rounded prescutellar one and a slightly transverse anterior one. Lateral pronotal margins regularly curved, anterior margin almost straight. Prehumeral pronotal carinae undeveloped. Elytra slender, 3.0–3.2 times as long as wide at humeral part, simply rounded and serrate apically. Anal sternite rounded. ♂: frontal pubescence rather dense and long; eyes rather large and frons narrow. ♀: frontal pubescence sparser and shorter; eyes smaller, and frons wider. Length: 8.3–13.0 mm.

Distribution. Not in Denmark or Finland. Sweden. Sk., Bl., Sm., Ög., Vg., and Upl. Norway: only VE and AAy on the southern coast. – Distributed from Great Britain through Europe to Baltic region of the USSR (Estonia, Lithuania, Latvia), Ukraine, Caucasus, and Transcaucasus; also North Africa.

72

Biology. The larva develops in and under the bark of standing or laying trunks and stumps of various *Quercus* species. Hellrigl (1978) also mentions *Fagus sylvatica* and *Castanea sativa*. The larval tunnels are flat and ususally horizontal. The development lasts two years. Pupation takes place in May, and the pupal chamber is made in thick bark. The adults appear in May-July on the host plants. Description of larva: Goureau (1843), Schiødte (1870), Alexeev (1960, 1961, 1981).

27. *Agrilus convexicollis convexicollis* Redtenbacher, 1849

Agrilus convexicollis Redtenbacher, 1849: 285.

A small and rather robust species. Dorsal side green-bronze or black-bronze, pronotum and head dark golden green with a brassy tinge; lateral pronotal margins blue-green. Ventral side black. Dorsal pubescence white, short and dense. Frons and vertex very convex, slightly grooved medially. Pronotum 1.5 times as wide as long, very convex and microsculptured. Lateral pronotal margins almost straight, anterior pronotal margin lobate in middle. Prehumeral pronotal carinae long and sharp, slightly convergent anteriorly, and reaching to middle of pronotum. Elytra 2.9 times as long as wide at base, rather convex, simply rounded, and finely serrate apically. Apical margin of anal sternite slightly incurved. Anterior prosternal lobe arched, without a notch. Prosternal process slightly depressed, mesosternum grooved. ♂: frons green, legs and antennae blue-green; frontal and prosternal pubescence rather long and dense. ♀: frons bronze, legs and antennae black; frontal and prosternal pubescence shorter and sparser. Length: 3.5–5.5 mm.

Distribution. Only in Sweden. Öl. and Gtl. – The nominate ssp. ranges from France to Latvia, Ukraine and the Crimea; and from the Balkans to South Scandinavia.

Biology. The larva develops in thin branches and twigs of *Fraxinus excelsior, F. ornus, F. oxyphylla, Ligustrum vulgare* and *Syringa* sp. It pupates in May or June, and the adult appears in June-July on the host plants. Description of larva: Alexeev (1981).

28. *Agrilus cyanescens* Ratzeburg, 1837.
Figs. 85, 96.

Buprestis caerulea Rossi, 1790: 407 (praeocc. Thunberg, 1789).
Buprestis amethystinus Schoenherr, 1817: 258 (praeocc. Olivier, 1790).
Buprestis cyaneus Lacordaire, 1835: 612 (praeocc. Rossi, 1790).
Agrilus cyanescens Ratzeburg, 1837: 62.
Agrilus amabilis Gory & Laporte de Castelnau, 1841: 52.
Agrilus sulcaticeps Abeille de Perrin, 1870: 79.
Agrilus bidentulus Ganglbauer, 1889: 31.

Agrilus fissifrons Abeille de Perrin, 1897: 3 (nec Fairmaire).
Agrilus kyselyi Obenberger, 1924: 50.

A rather short and robust species. Dorsal side blue-green or blue, in rare cases with a golden green pronotum. Ventral side blue or black-blue. Pubescence extremely short, almost indistinct, both dorsally and ventrally. Frons, and especially vertex, grooved medially, vertex with concentric lines of small punctures. Eyes very small. Pronotum 1.6 times as wide as long, with a lobate anterior margin and regularly curved lateral margins. Medial pronotal groove indistinct, lateral depressions small, but deep. Prehumeral pronotal carinae obtuse and indistinct. Elytra robust, 2.6–2.8 times as long as wide at base, somewhat broadened at posterior third. Apex of each elytron very slighly serrate and rounded. Anal sternite with an apical incurvation, anterior prosternal lobe notched medially. ♂: antennae somewhat longer, their medial segments rather serrate; eyes rather large; last sternite very slightly serrate apically. ♀: antennae shorter, their medial segments less serrate (with obtuse outer angles); eyes smaller; last sternite smooth apically. Length: 4.5–7.5 mm.

Distribution. Denmark: SJ, EJ, WJ, rather frequent in SE parts of Jutland, very local in the western part. Not in Sweden, Norway, or Finland. – Range: Spain to Lithuania, Ukraine and Caucasus; Balcan Peninsula to southern Jutland; introduced to the U.S.A.

Biology. The larva develops in the trunks and thick branches of various *Lonicera* species. It bores undulating tunnels in the superficial parts of the wood. The pupation takes place in May-June, adults appearing in June–July on the leaves of the host plants. Description of larva: Alexeev (1960, 1961, 1981).

29. *Agrilus guerini* Lacordaire, 1835
Plate 2: 11. Fig. 66.

Agrilus guérini Lacordaire, 1835: 608.

A large, robust and posteriorly acuminate species. Entire body blue or black-blue, sometimes with a slight green tinge. Elytra with six tomentose spots; all laterotergites and lateral parts of sternites with small areas of white tomentum. Frons flat, vertex grooved medially, entire head with rather long white pubescence. Pronotum 1.3 times as wide as long, with wide and shallow prescutellar and lateral depressions. Prehumeral pronotal carinae rudimentary. Transverse scutellar carina feeble. Elytra 3 times as long as wide, margins subparallel, sharply acuminate in posterior third. Each elytron with a very marked outer apical spine; these spines are slender and divergent (Fig. 66). Prosternal process with subparallel margins, and with long white pubescence. Anal tergite with a sharp medial keel projecting beyond the apical margin of the tergite. Entire ventral side with short and sparse white pubescence; anal sternite with long bristles on apical margin. ♂: eyes rather large and convex; frontal and prosternal pubescence rather long and dense; sternites II and III slightly grooved medially; anal sternite slightly incurved apically; fore and mid tibiae with a small inner apical spine. ♀: eyes smaller, less convex; frontal and prosternal pubescence shorter and sparser; sternites II and III not grooved medially; anal sternite rounded apically; fore and mid tibiae simple. Length: 8.5–12.0 mm.

Distribution. Sweden: Sm., occurring in a limited area N of Kalmar, with the centre around Hornsö. Not in Denmark, Norway, or Finland. – Also Central Europe and Ukraine.

Biology. A very rare and relict species. The larva develops under the bark of live branches and trunks, 2–10 cm in diameter, of various *Salix* species, with a preference for *S. caprea*. The tunnel of the larva is undulating, about 60–80 cm long, and runs parallel with the axis of the branch. Pupation takes place in May-June in the superficial parts of the wood, the adult appearing in June-July on leaves of the host plant. Description of larva: Alexeev (1981).

Agrilus hyperici (Creutzer, 1799)

Buprestis hyperici Creutzer, 1799: 122.
Agrilus prasinus Mulsant, 1863: 17.
Agrilus modestulus Semenov, 1895: 247.
Agrilus elatus Méquignon, 1907: 120.
Agrilus sulcifer Bétis, 1914: 458 (nec Abeille).
Agrilus catalonica Pochon, 1963: 65.

A small, cylindrical and robust species. Entire body bronze or violet-bronze, ventral side usually somewhat darker. Both dorsal and ventral side with very fine, but distinct, white pubescence. Head large, eyes small. Frons and vertex slightly convex, indistinctly grooved medially. Vertex with concentric rows of fine punctures. Pronotum robust and convex, 1.6–1.7 times as wide as long. Medial pronotal groove divided into a distinct polygonal prescutellar depression and an almost indistinct anterior depression. Lateral pronotal margins feebly curved and indistinctly incurved before posterior angles. Lateral pronotal depressions absent, prehumeral pronotal carinae very indistinct, or totally absent. Elytra robust, subcylindrical, 2.7–2.8 times as long as wide at humeral part. Apex of elytra obtuse, with very fine, almost indistinct serrations. Each elytron separately rounded. Anal sternite rounded apically. ♂: prosternal pubescence rather long and dense; frons with greenish tinge. ♀: prosternal pubescence shorter and sparser; frons bronze. Length: 3.5–6.0 mm.

Distribution. Not in Denmark or Fennoscandia. – Throughout Europe to Baltic region of the USSR (Estonia, Lithuania), Ukraine and Caucasus, south to Italy. Introduced to U.S.A.

Biology. The larva develops in roots of *Hypericum perforatum* and *H. tetrapterum*. The development lasts one year, pupation taking place in June. The adults appear in June-August. Description of larva: Perris (1877), Alexeev (1981).

30. *Agrilus integerrimus* (Ratzeburg, 1839)
Figs. 76, 99.

Buprestis integerrimus Ratzeburg, 1839: 64.
Agrilus cupreus Redtenbacher, 1849: 286.

A large, robust and rather convex species. Dorsal side bronze with a green tinge, or copper-coloured, pronotum with a coppery-red sheen, rarely entire body brightly bronze. Ventral side bronze. Pubescence indistinct on entire body. Head large and convex, frons and vertex grooved medially. Eyes relatively small. Vertex with fine concentric grooves (Fig. 76) composed of small and fine punctures. Pronotum very wide, convex, 1.8 times as wide as long. Lateral pronotal margins regularly rounded, slightly incurved before posterior angles. Lateral pronotal depressions slight, medial pronotal groove well developed. Prehumeral pronotal carinae indistinct. Elytra robust and slightly flattened, 2.7 times as long as wide at base. Each elytron rounded separately, and slightly serrate. Anterior prosternal lobe slightly bent, anal sternite rounded apically. ♂: eyes rather large; frons green-bronze. ♀: eyes smaller; frons bronze. Length: 5.0–7.5 mm.

Distribution. Not in Denmark, Sweden, or Norway. Finland: Ab, Ta, and Sa; also Kr in the USSR. – Spain to the Baltic region of the USSR (Estonia, Lithuania, Latvia), Ukraine and Caucasus, south to Sicily.

Biology. The larva develops in the wood of trunks and roots of *Daphne mezereum, D. laureola* and *D.gnidium*. The larva starts its tunnelling under the bark and continues (from the 2nd or 3rd instar) in the wood. The pupal chamber is situated in the wood. Pupation takes place in June, the adult appearing June-August on twigs and leaves of the host plants. The development lasts one year. Description of larva: Alexeev (1981).

31. *Agrilus laticornis* (Illiger, 1803)
Figs. 71, 73.

Buprestis laticornis Illiger, 1803: 243.
Agrilus scaberrimus Ratzeburg, 1837: 55.
Agrilus aceris Chevrolat, 1837: 5.
Agrilus laticollis Kiesenwetter, 1857: 142.
Agrilus asperrimus Marseul, 1865: 492.

A small and slender species. Dorsal side olive-green or green-bronze, pronotum usually black-green. Ventral side black-bronze. Dorsal and ventral pubescence sparse and short, almost indistinct, except for on the frons. Frons and vertex convex, vertex with longitudinal lines of fine punctures. Eyes large, but not projecting beyond outline of head. Pronotum 1.5–1.6 times as wide as long, with a slight medial groove. The maximum pronotal width at anterior margin. Lateral pronotal margins slightly incurved before posterior angles. Prehumeral pronotal carinae well developed, reaching almost to middle of pronotum. Elytra 3 times as long as wide at humeral level, very finely serrate and rounded together at apex. Anal sternite slightly incurved apically, anterior prosternal lobe notched medially. Prosternum flat, mesosternum not grooved medially. Prosternal process (Fig. 71) widening behind fore coxae. ♂: frons green; eyes rather large; antennal segments 4–10 very enlarged, 1.6 times as wide as long; frontal pubescence rather long and dense. ♀: frons bronze; eyes smaller; antennal segments 4–10 moderately enlarged, only 1.1 times as wide as long, with sharp outer angles; frontal pubescence shorter and sparser. Length: 4.3–6.2 mm.

Distribution. Denmark: SJ, EJ, F, LFM, NEZ, especially found in deciduous forests of C Zealand and Lolland. Sweden: in most districts from Sk. north to Sdm. and Vstm. Norway: only TEy and AAy on the southern coast. Finland: only N (Pernå and Orimattila). Not in Karelian part of the USSR. – Throughout the whole of Europe, from Great Britain eastwards to Asia Minor and Caucasus; North Africa to C Fennoscandia. Recorded from Lithuania and Latvia.

Biology. The larva develops under the bark of thin branches and twigs of various *Quercus* species and *Corylus avellana*. The development lasts one year, and pupation takes place in May-June. The adults can be seen on oak-leaves in May-June. Larva undescribed.

32. *Agrilus mendax* Mannerheim, 1837
   Figs. 63, 83, 88, 94.

Agrilus mendax Mannerheim, 1837: 111.
Agrilus faldermanni Gory & Laporte de Castelnau, 1841. 42.

A large, robust and posteriorly acuminate species. Dorsal side brass or copper-coloured, pronotum bronze with a green sheen. In rare cases elytra golden green with a coppery tinge and pronotum blue-green. Ventral side black-green with a blue tinge. Pubescence of entire body, including the frons, very short and sparse, almost indistinct. Head large, frons almost flat, vertex convex. Eyes small and flat. Vertex with fine longitudinal grooves composed of small and fine punctures. Pronotum convex, 1.4 times as wide as long, regularly rounded laterally and slightly incurved before posterior angles. The maximum width of pronotum at anterior third. Lateral pronotal depressions small and shallow. Medial pronotal groove divided into a shallow transverse anterior depression and a small and shallow, oval prescutellar depression. Prehumeral pronotal carinae short, but very elevated and arched. Scutellum very wide, almost trapezoidal. Elytra slightly flattened, 3 times as long as wide at base and sharply acuminate in posterior third. Sutural margins slightly divergent at apex, with fine indistinct serrations. Each elytron slightly curved outwards apically, as in *A. viridis* and *A. suvorovi populneus*. Anal sternite rounded apically, anterior prosternal lobe deeply notched. ♂: eyes rather large; tibiae slightly incurved on inner margins. ♀: eyes somewhat smaller; tibiae simple, not incurved. Length: 10.0–12.0 mm.

Distribution. Not in Denmark or Norway. In Sweden only in Dlr.: Bjursås N Falun, 1 specimen 1917 (E. Klefbeck); E of Leksand, several specimens 1980–81 (K. E. Forsman, S. Lundberg). Finland: in some southern districts, Ab, N, St, Ta, and Sa. Also Vib and Kr in the USSR. – Central and NE Europe, including also Estonia, Lithuania, and Latvia.

Biology. The larva develops under the bark, or partially in the wood, of branches and trunks of live *Sorbus aucuparia* and *S. aria*. The tunnels are strongly undulating and concentrated to about 10 cm of the trunk, but several attacks may appear along the same trunk. New attacks are situated at the margins of the attacks from the preceding year. Pupation takes place in May in the wood, adults appearing in June-July on leaves and trunks of the host trees. The development lasts one year, according to Lundberg (in litt.) two years in Sweden, Dalarne. Description of larva: Kangas (1947), Alexeev (1958, 1960, 1961, 1981).

33. *Agrilus olivicolor* Kiesenwetter, 1857
Fig. 70.

Agrilus olivaceus Ratzeburg, 1839: 61 (praeocc. Gyllenhal, 1808).
Agrilus olivicolor Kiesenwetter, 1857: 135.

A small and slender species. Entire body dark olive-green or brownish green, pronotum and head with a bronze tinge. Dorsal white pubescence confined to a band on either side of the elytral suture. Ventral pubescence extremely fine and almost indistinct, except on the male prosternum, which is clothed with white hairs at the centre. Head small, frons and vertex convex and grooved medially. Vertex with traces of longitudinal lines composed of very fine punctures. Eyes large, but not projecting beyond outline of head. Pronotum convex, subcylindrical, 1.3–1.4 times as wide as long, distinctly grooved medially. Lateral pronotal depressions small and shallow. Lateral pronotal margins almost straight, very slightly incurved before posterior angles. Prehumeral pronotal carinae well developed, sharp and long, reaching almost to middle of pronotum. Elytra slender, 3.3 times as long as wide at humeral level, slightly widening at posterior third. Elytral apex slender, only very feebly serrate. Sutural angle almost indistinct. Anal sternite incurved apically, anterior prosternal lobe arcuate. Prosternum flat, mesosternum not grooved. ♂: eyes rather large; abdominal sternite II with two small warts (Fig. 70); prosternum with an area of white hairs. ♀: eyes smaller; abdominal sternite II without warts; prosternum bare. Length: 4.0–5.5 mm.

Distribution. Sweden: Sk., Bl., Sm., Öl., Ög., Vg., and Upl. Norway: AK, Bø. Not in Denmark or Finland, but recorded from Vib in the USSR. – France to Siberia, south to Italy.

Biology. The larva develops under the bark of branches and twigs of live *Corylus avellana, Carpinus betulus* and *Fagus sylvatica*. The development lasts one year, pupation taking place in May-June, and the adult appearing in May-August. Larva undescribed.

34. *Agrilus pratensis pratensis* (Ratzeburg, 1839)
Plate 2: 17.

Buprestis linearis Paykull, 1799: 226 (praeocc. Panzer, 1789).
Buprestis pratensis Ratzeburg, 1839: 63.
Agrilus roberti Chevrolat, 1838: 89.

A small and slender species. Frons green or bronze, vertex black. Pronotum golden bronze or golden green, elytra olive-green, blue or blue-violet. In rare cases elytra bronze, or entire body blue with a greenish pronotum. Ventral side green-black. Pubescence of entire body indistinct except for on frons, genae and prosternum, which have a short, but dense, white pubescence. Frons and vertex convex, vertex with fine linear puncturation. Eyes small, but slightly projecting beyond outline of head, especially ind the male. Pronotum convex, 1.5–1.6 times as wide as long, with large and deep lateral depressions. Medial pronotal groove divided into a large and deep, slightly elongate, prescutellar depression, and a smaller, somewhat transverse, anterior depression. Lateral pronotal margins angulate at middle, maximum pronotal width at anterior margin. Prehumeral pronotal carinae short, sharp and straight. Elytra 3.0–3.1 times as long as wide at base, only slightly convex. Each elytron finely serrate and simply rounded at apex. Anal sternite rounded apically, anterior prosternal lobe arcuate. ♂: eyes slightly projecting beyond outline of head; fore and mid tibiae slightly incurved on inner margin; frontal and prosternal pubescence rather well developed. ♀: eyes smaller, not projecting beyond outline of head; fore and mid tibiae simple; frontal and prosternal pubescence shorter. Length: 4.0–6.5 mm.

Distribution. Not in Denmark. Sweden: in many districts from Sk. north to Dlr. and Hls. Norway: only TEy (Kragerø). Finland: Ab, N, Ka, Ta, Sa, Tb. USSR: Kr. – Europe to Siberia and Asia Minor, south to Sicily; not in Great Britain, Spain or Portugal.

Biology. The larva undertakes its development under the bark of branches of various *Populus* species. It pupates in May-June, the adult appears in June-July. The development lasts one year. Larva undescribed.

35. *Agrilus pseudocyaneus pseudocyaneus* Kiesenwetter, 1857

Agrilus pseudocyaneus Kiesenwetter, 1857: 150.

A small and robust species. Entire body blue or blue-violet. Legs and sternites black with a blue tinge. Pubescence of whole body quite indistinct. Head relatively large, convex, frons slightly depressed medially. Vertex with concentric circles of very fine punctures. Eyes small, not projecting beyond outline of head. Pronotum convex, distinctly grooved medially, 1.5–1.6 times as wide as long. Anterior pronotal margin lobate medially, lateral margins almost parallel-sided in anterior half, slightly incurved before posterior angles. Lateral pronotal depressions very deep and large. Prehumeral pronotal carinae very feeble and short, converging anteriorly. Entire pronotum with very fine microsculpture. Elytra rather robust, 3.5 times as long as wide at base, somewhat widened at posterior third. Elytra slightly serrate apically, each elytron separately rounded. Anal sternite rounded apically, anterior prosternal lobe arcuate. ♂: tibiae slightly incurved on inner margin; medial antennal segments triangular, with sharp outer angles; prosternum with extremely short white hairs. ♀: tibiae simple; medial antennal segments enlarged, with obtuse outer angles; prosternum glabrous. Length: 4.0–5.5 mm.

Distribution. Not in Denmark, Sweden, or Norway. Finland: only Sb (Vehmersalmi); USSR: Vib and Kr.; also Estonia, Lithuania, and Latvia. – Central and southern Europe.

Biology. The larva develops under the bark of thin branches and twigs of *Salix caprea* and *S. viminalis*. The development lasts one year, and pupation takes place in May-June. The adults are seen in June-July on leaves and twigs of the host plants. Larva undescribed.

36. *Agrilus subauratus* Gebler, 1833
Plate 2: 16. Figs. 58, 68, 87, 92.

Buprestis subauratus Gebler, 1833: 277.
Agrilus coryli Ratzeburg, 1837: 55.
Agrilus auripennis Gory & Laporte de Castelnau, 1841: 46.

A large, robust and cylindrical species. Dorsal side golden green, head and pronotum blue, or entire body blue; rarely elytra golden red or red, head and pronotum blue green. Ventral side blue or blue-green. Dorsal pubescence quite indistinct. Head relatively small, frons and vertex slightly grooved medially. Eyes very small, vertex with concentric lines of small punctures. Pronotum wide, 1.6–1.8 times as wide as long, with well developed and deep lateral depressions. Lateral pronotal margins somewhat lobate at middle, slightly incurved before posterior angles. Prehumeral pronotal carinae indistinct or not developed at all. Scutellum without a transverse carina (usually present in *Agrilus*). Elytra 3 times as long as wide at base, widening at posterior third, simply rounded and very slightly serrate apically. Apical margin of anal sternite slightly incurved. Prosternal process subparallel, pointed apically. ♂: prosternal pubescence well developed, dense. ♀: prosternal pubescence sparse and short, almost indistinct. Length: 6.5–10.0 mm.

Distribution. Not in Denmark or Norway. Sweden: older finds in Ög., Sdm., and Upl.; in recent years found at Dalälven, Upl., breeding in *Salix cinerea* (Baranowski, 1980), and in Sm., Rugstorp, 3 specimens (W. Kronblad & F. Olsson in litt.). Finland: Ah, N, Ka, St, Ta, Kb, USSR: Vib, Kr. – France to Siberia, south to the Balkans.

Biology. The larva develops in live wood of various *Salix* species, preferring *Salix caprea*. The tunnels are about 100 cm long and run parallel with the axis of trunk or branch. Pupation takes place in May, the pupal chamber being situated in the superficial parts of the wood. The adults appear in June-July on the host plants. Description of larva: Alexeev (1960, 1961, 1981).

37. *Agrilus sulcicollis* Lacordaire, 1835
   Figs. 1, 2, 57, 59, 61, 72, 75, 82.

Buprestis cyanea Rossi, 1790: 189 (praeocc. Fabricius, 1775).
Agrilus sulcicollis Lacordaire, 1835: 614.
Agrilus tenuis Ratzeburg, 1837: 53.
Agrilus sahlbergi Mannerheim, 1837: 113.
Agrilus viridis Gory & Laporte de Castelnau, 1844: 48 (nec Linné).

A large, robust and cylindrical species. Dorsal side green, brown-green, blue-green, blue or violet. Ventral side black with a green or bronze tinge. Pubescence of entire body short and indistinct except for on the frons and prosternum, which bears a rather long and dense white pubescence. Frons almost flat, vertex slightly grooved medially. Eyes large, but not projecting beyond outline of head. Vertex with fine longitudinal grooves composed of fine punctures. Pronotum 1.6 times as wide as long, with a wide and deep medial groove. Maximum pronotal width at anterior margin. Lateral pronotal depressions large and deep. Prehumeral pronotal carinae feeble, short and arched, sometimes almost indistinct. Elytra cylindrical, 3.5 times as long as wide at humeral level. Apex of elytra rounded, very obtuse and indistinctly serrate. Anal sternite incurved apically, anterior prosternal lobe arched. Prosternum flat, mesosternum not grooved medially. ♂: eyes large; frontal and prosternal pubescence rather long and dense; tibiae incurved on distal part of inner margin, with an inner apical tooth; sternite II with two sharp warts, as in *angustulus;* aedeagus (Fig. 75) robust, very asymmetric. ♀: eyes smaller; frontal and prosternal pubescence shorter and sparser; tibiae simple; sternite II without warts. Length: 6.0–8.5 mm.

Distribution. Denmark: very rare, the first breeding records date from about 1960, since seemingly spreading; LFM, NWZ, NEZ. Sweden: Sk., Bl., Hall., Sm., Öl., Ög., Dlsl., Sdm., Upl., Vstm., Gstr. Norway: AK, VE, TEy, AAy. Finland: Ab, N, St, Ta. USSR: Vib. – Throughout whole Europe, excepting Great Britain, south to Italy, east to Caucasus.

Biology. The larva develops in or under bark of live *Quercus* species. The development lasts 1–2 years, pupation taking place in May-June, usually in the bark. The adults can be met with in May-July on leaves and branches of the host plants. Description of larva: Palm (1962), Alexeev (1981).

38. *Agrilus suvorovi populneus* Schaefer, 1946
   Figs. 79, 80.

Agrilus viridis var. *populnea* Schaefer, 1946: 73.
Agrilus viridis ab. *cyanophila* Schaefer, 1946: 73.
Agrilus suvorovi var. *pseudofagi* Obenberger, 1955: 45.

A large, long and slender species resembling large specimens of *A. viridis*. Entire body olive-green with a bronze lustre, blue-green, or blue. Pubescence of dorsal side indistinct, of ventral side white, very fine and sparse. Frons and prosternum with distinct, dense but short, white pubescence. Head relatively large, eyes small. Frons and vertex convex. Vertex sometimes slightly grooved medially, with longitudinal lines composed of small punctures. Pronotum enlarged anteriorly, 1.5–1.6 times as wide as long; its maximum width at anterior quarter. Lateral pronotal margins regularly curved and slightly incurved before posterior angles. Lateral pronotal depressions large and deep; medial groove divided into a large, elongate prescutellar depression, and a smaller, transverse, anterior depression. Prehumeral pronotal carinae straight and short, somewhat obtuse. Elytra long and slender, somewhat flattened, 3.6–3.7 times as long as wide at base. Apex of elytra distinctly serrate, slender (Fig. 80); sutural margins somewhat divergent apically and each elytron slightly outwardly bent. Anal sternite rounded apically, anterior prosternal lobe arcuate. ♂: frontal and prosternal pubescence rather well developed; medial antennal segments wider, with sharp outer angle. ♀: frontal and prosternal pubescence less developed; medial antennal segments narrower, with obtuse outer angle. Length: 7.0–10.5 mm.

Distribution. Sweden: Bl., Sm., Ög., Dlsl., Upl.. Norway: TEy and AAy. Not in Denmark or in East Fennoscandia. – France to Ukraine, south to Italy. The nominate ssp. was described from Siberia.

Biology. The larva develops under the bark of live branches and thin trunks of *Populus tremula*. The development lasts one year, pupation taking place in June-July. The biology of this species was completely studied by Arru (1961–62), who also described the larva.

39. *Agrilus viridis viridis* (Linné, 1758)
   Figs. 62, 69, 77, 78, 84, 93.

Buprestis viridis Linné, 1758: 410.
Mordella serraticornis Scopoli, 1763: 61.
Mordella rosacea Scopoli, 1763: 61.
Buprestis linearis Panzer, 1789: 101.
Buprestis filiformis Herbst, 1801: 313.
Buprestis fagi Ratzeburg, 1837: 56.
Buprestis nociva Ratzeburg, 1837: 56.
Agrilus distinguendus Gory & Laporte de Castelnau, 1841: 44.
Agrilus viridipennis Gory & Laporte de Castelnau, 1841: 45.
Agrilus aubei Gory & Laporte de Castelnau, 1841: 45.
Agrilus capreae Chevrolat, 1838. 56.
Agrilus littlei Curtis, 1840: 365.
Agrilus quercinus Redtenbacher, 1849: 287.
Agrilus bicolor Redtenbacher, 1849: 287.

A large, long and slender species. Body green, olive-green, golden green, bronze, bronze-green, blue-green, blue, violet or black. Dorsal pubescence extremely short, almost indistinct. Frontal pubescence and pubescence of ventral side white and short, but rather dense. Frons and vertex slightly convex, vertex with a fine medial groove and longitudinal rows of fine punctures (Fig. 77). Eyes not projecting beyond outline of head. Pronotum 1.5–1.6 times as wide as long, lateral margins almost straight or regularly curved, with small and shallow lateral depressions. Lateral pronotal margins usually slightly incurved before posterior angles. Medial pronotal groove divided into a shallow anterior transverse depression and an elongate prescutellar depression. Prehumeral pronotal carinae short, slightly bent and usually obtuse. Elytra about 3.5 times as long as wide at base, slightly widened at posterior third, and distinctly serrate apically. Apical part of elytra slender, elongate, and slightly outwardly bent. Anal sternite rounded apically (Fig. 69), anterior prosternal lobe arcuate (Fig. 84). ♂: eyes rather large; frontal and prosternal pubescence rather dense and long; medial antennal segments sharply triangular, segments 10 and 11 approximately of the same shape and size. ♀: eyes smaller; frontal and prosternal pubescence sparser and shorter; medial antennal segments with obtuse outer angles, segment 11 smaller and more slender than segment 10. Length: 4.5–11.0 mm.

Distribution. Denmark: in all districts except for NWJ and F; rather seldom, most abundant in boggy areas of Jutland. Sweden: recorded from all districts except for G. Sand. and Med. Also in Norway known from most districts, but apparently absent from the districts (R, HO, SF) along the southwestern coast. Finland: in all districts except for LkE, Le and Li in the north. USSR: Vib and Kr. – The nominate ssp. is distributed from Great Britain to Siberia, south to North Africa. Several ssp. are described from the East Palaearctic Region.

Biology. An extremely polyphagous species. The larva develops under the bark of branches of various species of *Salix, Alnus, Betula, Carpinus, Corylus, Acer, Fagus*, and *Tilia*. The development lasts one year, the pupation taking place in May-June. The pupal chamber is made in the wood. The adults are seen in May-September and can be collected from leaves of willow and birch. Description of larva: Kangas (1947), Alexeev (1960, 1961, 1981).

Note. *A viridis* is one of the most variable species of buprestid beetles. It shows an extreme variation not only in size and coloration, but also in structure and proportions of the pronotum and in shape of the elytra. Also the aedeagus varies according to the various host plants (Alexeev, 1969). The *viridis* species-group (*viridis, aurichalceus, suvorovi* etc.) is a phylogenetically young group, which is in a state of radiation. This has caused the adaptibility to various host plants and the extreme variability associated with this phenomenon. Many species, subspecies and other lower taxa have been described in this group, but they are usually only to be treated as ecological forms without firm taxonomic value.

Tribe Aphanisticini J. du Val, 1859

Small-sized, slender species; scutellum without transverse keel; basal segment of hind tarsus only slightly longer than segment 2 (Fig. 60); pronotum with simple lateral edge, and with simple, fine sparse puncturation; prosternum and femora with grooves for the reception of the antennae and the tibiae; frons with shallow, but wide, medial groove (Fig. 56); tarsi with only one claw.

Larva poorly known. Prothorax without grooves; anal segment simple, without sclerotized spines.

Genus *Aphanisticus* Latreille, 1810

Aphanisticus Latreille, 1810: 169.
  Type-species: *Buprestis emarginatus* Olivier, 1790.

Goniophtalma Chevrolat, 1837: 106.

Small to very small, elongate and subcylindrical species, with body black or dark bronze. Both lateral parts of mesosternum reduced, almost indistinct. Hind coxae outwardly enlarged. Scutellum very small, triangular, without a transverse keel. Basal segment of hind tarsi only slightly longer than segment 2 (Fig. 60), tarsi with only one claw. Pronotum with simple lateral margins. Eyes small, reniform, not reaching to anterior pronotal margin. Antennal segments 1 and 2 enlarged and swollen. Prosternum with grooves for the reception of the antennae. Femora with narrow grooves for the reception of the tibiae. Head always much narrower than pronotum.

Larva: sclerotized anal spines as well as pronotal and prosternal sclerotized grooves absent.

There are about 300 species in the world fauna, most af them are found in the Oriental and Afrotropical Regions. The palaearctic fauna comprises 15 species, of which 6 are known to occur in Europe.

Key to species of *Aphanisticus*, adults

1. A short and robust species (Fig. 100); elytra only twice as long as wide; pronotum 1.2 times as wide as long, with a very indistinct transverse depression; medial frontal depression reaches to anterior margin of pronotum; basal half of elytra with punctures forming indistinct longitudinal rows; apical half of elytra irregularly punctured; 2.2–3.0 mm .... 40. *pusillus* (Olivier)
   – An elongate and slender species, elytra 3 times as long as wide; pronotum about as long as wide; with two transverse, shallow depressions; medial frontal depression not reaching anterior margin of pronotum; elytra finely punctured, punctures form-

84

ing longitudinal rows which almost reach the elytral apex;
2.5–3.8 mm .... *emarginatus* (Olivier)

40. *Aphanisticus pusillus* (Olivier, 1790)
Figs. 60, 100.

Buprestis pusilla Olivier, 1790: 91.
Buprestis emarginatus Fallén, 1802: 11 (nec Olivier).
Buprestis lineola Germar, 1834: 10.

A very small, but rather robust species. Entire body completely black and lustrous, sometimes with a very slight bronze tinge. Frons and vertex with a deep and wide groove, which reaches the anterior pronotal margin. Temporal parts of head almost flat. Pronotum smooth and convex, 1.2–1.3 times as wide as long, with only a very slight and indistinct transverse depression at anterior third. Lateral pronotal margins regularly rounded and slightly incurved before prosterior angles. Lateral pronotal depressions wide and very enlarged posteriorly. Elytra convex and robust, only 2.0–2.2 times as long as wide at base. Apical margins of elytra without serrations, only basal half of elytra with feeble rows of punctures. Anal sternite rounded apically. Sexual dimorphism almost non-existant. ♂: tergite VIII arcuated apically. ♀: tergite VIII also arcuated but somewhat elongate apically. Length: 2.2–3.0 mm.

Distribution. Denmark: in all districts except FJ, WJ, and NWJ, most frequent near the coasts of the eastern islands. In Sweden only in the south: Sk., Bl., Hall., Sm., Öl., Vg., and Boh. No records from Norway or Finland. – Throughout Europe from Great

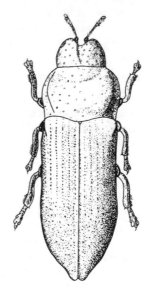

Fig. 100. Habitus of *Aphanisticus pusillus* (Oliv.), 2.2–3.0 mm.

Britain to Baltic region of the USSR (Estonia, Lithuania, Latvia) and Ukraine, south to North Africa.

Biology. The larva develops in stalks of various *Juncus* species, probably also in some *Carex*. Details of the life history are unknown, and the larva is as yet undescribed.

Aphanisticus emarginatus (Olivier, 1790)
Fig. 56.

Buprestis emarginata Olivier, 1790: 84.
Aphanisticus coriaceus Rey, 1891: 27.
Aphanisticus angustulus Ragusa, 1893: 300.

A small, subcylindrical and elongate species. Entire body black, rarely with a brassy tinge. Head relatively large, eyes small. Frons and vertex deeply and widely grooved, temporal part of head convex. Frontal groove not reaching anterior pronotal margin. Pronotum 1.1 times as wide as long, with two distinct transverse depressions: a rather low anterior one and a deeper postmedial one. Lateral pronotal margins regularly rounded at anterior half, straight but convergent, at posterior half. Lateral pronotal depression wide, mainly in posterior part. Elytra subcylindrical, 3.2 times as long as wide at base, sometimes broadened at posterior third. Apical part of elytra slender. Elytra with rows of fine punctures; these rows reach the posterior quarter of elytra. Anal sternite rounded apically. Sexual dimorphism only very slight. ♂: tergite VIII straight apically. ♀: tergite VIII feebly incurved apically. Length: 2.5–3.8 mm.

Distribution. So far not recorded from Denmark or Fennoscandia. – Baltic region of the USSR: Estonia, Lithuania, Latvia (Obenberger, 1937; Schaefer, 1949), but see comments below. Throughout Europe from Great Britain to Caucasus and Asia Minor, and from North Africa to Central Europe.

Biology. The larva develops in stalks of *Juncus obtusiflorus* and *J. articulatus*. The life history and larva are unknown.

Note. I have not seen specimens of *emarginatus* originating from northern Europe, including Fennoscandia. Obenberger's (1937) and Schaefer's (1949) records of this species from the Baltic region of the USSR probably refer to *pusillus* (Oliv.). I think the occurrence of *emarginatus* in Fennoscandia is very improbable.

SUBFAMILY TRACHYINAE

Small or very small, triangular or subtriangular species (Figs. 101, 108); frons with deep, punctiform pits above clypeal suture; pronotum at least 3 times as wide as long; development in leaf parenchyme.
Larva: abdominal segments very enlarged, laterally with spheric ampullae (Fig. 10); microspinules on entire body dark; prothorax narrower than mesothorax; pronotum,

prosternum and sometimes also abdominal segments with dark, sclerotized plates dorsally and ventrally. Only one tribe in the European fauna.

Tribe Trachyini Gory & Laporte de Castelnau, 1841

Key to genera of Trachyinae, adults

1 Elytron with a smooth and sharp carina reaching from humeral swelling to elytral apex (Fig. 108); prosternum with a narrow, medially notched, anterior lobe; scutellum triangular, well developed (Fig. 108); pronotum with deep punctiform depressions at anterior angles .... *Habroloma* Thomson (p. 91)
- Elytron smooth, without carinae (Fig. 101); prosternum without an anterior lobe; scutellum extremely small, almost invisible; pronotum without depressions at anterior angles ..... *Trachys* Fabricius (p. 87)

Key to genera of Trachyinae, larvae

1 Only pronotum and prosternum with sclerotized plates; abdomen composed of 11 segments; abdominal segments 2-7 equally wide, without lateral pubescence .... *Habroloma* Thomson (p. 91)
- All thoracic and abdominal segments with dorsal and ventral sclerotized plates (Fig. 105); abdomen 10-segmented; abdomen narrowed from segment 2 onwards; abdominal segments with lateral pubescence (Fig. 10) .... *Trachys* Fabricius (p. 87)

Genus *Trachys* Fabricius, 1801

Trachys Fabricius, 1801: 218.
   Type-species: *Buprestis pygmaea* Fabricius, 1787
Phytotera Gistl, 1856: 366.

Small or very small, triangular and convex species. Coloration varying from black to golden-bronze or blue, sometimes head and pronotum differently coloured, or elytra with transverse spots, stripes or tufts of rigid hairs. Frons always with two deep punctiform pits above the clypeal suture. Antennae short, with the two basal segments somewhat enlarged. Pronotum short and wide, 4-5 times as wide as long. Scutellum punctiform, almost invisible. Prosternum without an anterior lobe, prosternal process always margined. Elytra without lateral carinae.
   Larva: thoracic and abdominal segments with dorsal and ventral sclerotized plates (Figs. 10, 105); abdomen 10-segmented and narrowing from segment 2 onwards; abdominal segments with pubescence laterally.
   A large genus comprising about 500 species in the Old World. There are about 80 palaearctic species, of which 12 are known to occur in Europe.

Key to species of *Trachys*, adults

1   Elytra with projecting humeral swellings (Fig. 101), and with a shallow depression between suture and humeral swelling; elytra black, sometimes with a slight violet tinge, and with 4 narrow transverse zigzag pubescent bands; pronotum black with bronze sheen; prosternal process as in Fig. 103; larger species: 2.75–3.50 mm .... 41. *m. minutus* (Linné)

–   Elytra convex, without projecting humeral swellings or depressions; elytra without hairs or with a sparse irregular white pubescence 2

2 (1) Elytra with an irregular and sparse pubescence, which is never concentrated into transverse bands; vertex flat; bronze-black species; prosternal process as in Fig. 104; 1.5–2.5 mm .... 43. *scrobiculatus* Kiesenwetter

–   Elytra without hairs; slightly bicolorous species: elytra black with a blue tinge, head and pronotum dark bronze; prosternal process as in Fig. 102; 2.0–2.3 mm .... 42. *troglodytes* Gyllenhal

Key to species of *Trachys*, larvae

1   Prosternal sclerotized plate enlarged anteriorly (Fig. 106); host plants: *Salix, Betula, Ulmus, Corylus, Sorbus* .... 41. *minutus* (Linné)

–   Prosternal sclerotized plate narrowed anteriorly (Fig. 107); host plants: *Stachys recta, Mentha* .... 43. *scrobiculatus* Kiesenwetter

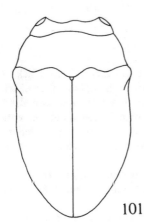

102   103   104

101

Fig. 101. Outline of *Trachys minutus* (L.).
Figs. 102–104. Prosternal process of adult of 102: *Trachys troglodytes* Gyll; 103: *T. minutus* (L.); 104: *T. scrobiculatus* Kiesw.

88

41. *Trachys minutus minutus* (Linné, 1758)
Plate 2: 13. Figs. 101, 103, 105, 106.

Buprestis minutus Linné, 1758: 410.
Trachys supraviolacea Thomson, 1864: 41.
Trachys mandjuricus Obenberger, 1917: 217.

A small and robust species. Entire body black or violet-black, pronotum with a slight bronze sheen. Pubescence short and sparse, but distinct, elytral pubescence concentrated into three transverse zigzag bands of white hairs. Head wide, frons slightly grooved medially. Eyes small, their inner margins slightly ridged. Pronotum very wide, its posterior margin distinctly margined. Discal part of pronotum with a simple, indistinct puncturation, posterior part with a shallow semicircular puncturation. Elytra wide and robust, 1.2 times as long as wide at base. Humeral swelling projecting and large. Elytra uneven with several shallow irregular depressions. Prosternal process enlarged at posterior third (Fig. 103). Sexual dimorphism non-existant. Length: 2.7–3.5 mm.

Distribution. Denmark: SJ, EJ, and LFM, rare. Sweden: in most districts from Sk. to T. Lpm. Norway: in the southern districts, north to On and SFi. Finland: in all districts in the south, north to Om and Ok. USSR: Vib and Kr. – Great Britain throughout Europe, east to Siberia, south to Italy.

Biology. The larva develops in leaf parenchyme of many *Salix* species, *Corylus avellana, Ulmus campestris,* and *Sorbus aria.* The larval development lasts 4–6 weeks. The adults hibernate in litter or in the grass layer. Oviposition takes place in April-May. Description of larva: Schiødte (1870), Hering (1951), Yano (1952).

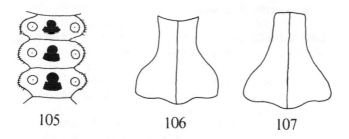

105   106   107

Fig. 105. Dorsal abd. sclerites of larva of *Trachys minutus* (L.).
Figs. 106, 107. Prosternal plate of larva of 106: *Trachys minutus* (L.) and 107: *T. scrobiculatus* Kiesw.

42. *Trachys troglodytes* Gyllenhal, 1817
Fig. 102.

Trachys troglodytes Gyllenhal, 1817: 125.
Trachys aenea Mannerheim, 1837: 122.

A small convex glabrous species. Head and pronotum black with a bronze tinge, elytra blue-black or blue. Ventral side black. Head wide, frons and vertex widely grooved. Pronotum very wide, with a fine microsculpture, and with wide, semicircular, almost horseshoe-shaped punctures on basal half. Lateral pronotal margins slightly arcuated. Elytra very convex, 1.2–1.3 times as long as wide at base, without humeral swellings. Basal half of elytra with oblique rows of large, shallow punctures; apical half with an irregular puncturation. Prosternal process very enlarged anteriorly (Fig. 102). Sexual dimorphism non-existant. Length: 2.0–2.3 mm.

Distribution. Denmark: EJ, LFM, NEZ; Sweden: Sk., Hall., Sm., Gtl., Vg., Boh., Nrk.; no records from Norway or Finland. – Great Britain throughout Europe to Estonia, Lithuania, Latvia, Ukraine and Caucasus, south to Italy.

Biology. The larva undertakes its development in leaf parenchyme of various species of *Scabiosa* and in *Knautia arvensis*. Also reared from *Succisa pratensis* (Huggert, 1967). Larva and life history undescribed.

43. *Trachys scrobiculatus* Kiesenwetter, 1857
Figs. 104, 107.

Trachys scrobiculatus Kiesenwetter, 1857: 171.
Trachys menthae Bedel, 1921: 225.
Trachys subacuminatus Pic, 1922: 29.
Trachys aeneus Théry, 1942: 191.
Trachys pumila auct. (nec Illiger, 1803; Castelnau et Gory, 1837; du Val, 1852).

A small, very convex and robust species. Entire body black with a slight bronze sheen, or dark bronze. Pubescence developed only on elytra. Hairs are yellowish and coarse, not concentrated into distinct transverse bands. Head large, frons depressed and grooved medially, vertex flat or very feebly grooved anteriorly. Inner margins of eyes not ridged. Pronotum very wide, regularly convex, with an extremely fine microsculpture (as on head), and with several very indistinct, wide, semicircular punctures along the posterior margin. Elytra wide, very convex, shiny, 1.1–1.2 times as long as wide at base. Humeral swellings very small, almost indistinct. Elytral puncturation irregular, consisting of large, but shallow and superficial, punctures. Prosternal process narrow, parallel-sided in anterior half, conspicuously widened posteriorly (Fig. 104). Sexual dimorphism indistinct. Length: 1.5–2.5 mm.

Distribution. Denmark: LFM (Knuthenborg Park), probably not a native species. Norway: AK, TEy. – Great Britain and Spain to Ukraine, south to North Africa.

90

Biology. The larva develops in leaf parenchyme of various *Mentha* species, according to Schaefer (1949) also in *Calamintha nepeta*. In Denmark reared from *Glechoma hederacea*. Description of larva: Schaefer (1949). Life history unknown.

Note. This widely distributed species has very often been confused with *Trachys pumila* (Illiger, 1803). Schaefer (1949) showed that *T. pumila* (Ill.) is a distinct species distributed only in Portugal and Spain. All specimens from Central and North Europe which I have seen, and which were labelled by various authors as *T. pumila* (Ill.), belonged to *T. scrobiculatus* Kiesw.

Genus *Habroloma* Thomson, 1864

Habroloma Thomson, 1864: 42.
  Type-species: *Buprestis nana* Paykull, 1799.

Small to very small, triangular and flattened species. Coloration very variable, rarely simply black, usually metallic bronze, often with spots or transverse bands composed of fine hairs, or with tufts of rigid bristles. This genus differs from *Trachys* by having a well developed anterior prosternal lobe and by the presence of a lateral elytral carina or keel reaching from humeral swelling to apex.

Larva: only pronotum and prosternum with dark, sclerotized plates; abdomen composed of 11 segments; abdominal segments 2-7 equally wide, without pubescence laterally.

The genus is distributed in the Palaearctic and Oriental Regions and comprises about 250 species, of which 25 are known to occur in the Palaearctic Region. Only two species occur in Europe. Kurosawa (1959) split off the subgenus *Parahabroloma* Kur. which contains only oriental species.

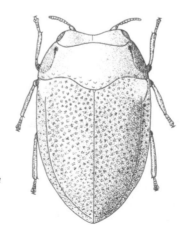

Fig. 108. Habitus of *Habroloma geranii* (Silfv.), 2.0-2.8 mm.

48. *Habroloma (Habroloma) geranii* (Silfverberg, 1977)
Fig. 108.

Buprestis minuta Rossi, 1790: 190 (praeocc. Linné, 1758).
Buprestis nana Paykull, 1799: 233 (praeocc. Gmelin, 1780).
Trachys geranii Silfverberg, 1977: 92.

A small, triangular and shiny species. Entire body completely black. Head relatively small, eyes with ridged inner margins. Frons widely depressed and grooved medially, vertex slightly grooved medially in anterior part and slightly convex in posterior part. Pronotum very wide, with somewhat S-shaped lateral margins, and with large and deep punctiform depressions at anterior angles. Pronotum without puncturation, only with a very fine and shiny microsculpture. Elytra somewhat flattened, shiny with coarse, quite irregular puncturation, 1.3 times as long as wide at base. Humeral swellings small, but distinct; each elytron with a sharp lateral carina reaching from humeral swelling to elytral apex. ♂: anal sternite slightly incurved apically, with an x-shaped structure. ♀: anal sternite rounded. Length: 2.0–2.8 mm.

Distribution. Denmark: only EJ (Helgenæs) and B (Hammeren). Sweden: in most districts from Sk. north to Dlr. and Gstr., also Ång.; Norway: only in the southeastern districts; Finland: Al, Ab, N, St, Ta, Sa, Tb, Kb, and Ok; USSR: Vib and Kr. – Throughout Europe, south to Italy, east to Siberia. Not in Great Britain.

Biology. The larva develops in leaf parenchyme of *Geranium sanguineum*. Oviposition takes place in the spring. The adults hibernate in litter and soil beneath the host plants. The habitat in Denmark is sun-exposed coastal slopes. Description of larva: Schaefer (1949).

Literature

Abeille de Perrin, E., 1870: Coléoptères nouveaux pour la France. – Abeille, Paris, 7: 47–48.
– 1897: Supplément à mes précédentes Notes sur les Buprestidae. – Revue Ent., 16: 33–37.
– 1900: Diagnoses de Coléoptères présumés nouveaux. – Bull. Acad. Marseille, 1900: 1–23.
Alexeev, A. V., 1958: Rjabinovaja zlatka *Agrilus mendax* Mannh. v Orechovo-Zuevskom rajone Moskovskoj oblasti. – Uc. zap. Orech.-Zuevsk. ped. inst., 11: 193–202.
– 1960: K morfologii i sistematike licinok nekotorych vidov zlatok roda *Agrilus* Curt. evropeiskoi casti SSSR. – Zool. Zh., 39: 1497–1510.
– 1961: Opredelitel zlatok roda *Agrilus* Curtis evropeiskoi casti SSSR. II. Opredelitel licinok. – Trudy po ekologii i sist. zhivotnych, 2: 3–21.
– 1964: O rozliciach mehzdu licinkami sinei sosnovoi (*Phaenops cyanea* F) i listvennicnoi (*P. guttulata* Gebl.) zlatok (Coleoptera, Buprestidae). – Ént. Obozr., 43: 644–650.
– 1981: Opredelitel licinok zlatok roda *Agrilus* Curtis (Coleoptera, Buprestidae) evropeiskoi casti SSSR. – Sbor. trudov zool. muz. MGU, 19: 65–84.

Alexeev, A. V., Zykov, I. B., 1979: Materialy po licinkam zlatok roda *Lampra* Lac. (Coleoptera, Buprestidae) Dalnego Vostoka i Vostocnoi Sibirii. – Sbor. "Zhuki Dal. Vostoka i Vost. Sibiriii", Vladivostok: 140–149.

Arru, G. M., 1961–62: *Agrilus suvorovi populneus* Schaef. (Coleoptera, Buprestidae) daunosa ai pioppi nell' Italia Settentrionale. – Boll. Zool. agr. Bachic., II, 4: 159–287.

Baranowski, R., 1980: Några bidrag till kännedomen om coleopter-faunan vid nedre Dalälven. – Ent. Tidsskr., 101: 29–42.

Baudon, A., 1968: Catalogue commenté des Buprestidae récoltés au Laos. Ministére des l'Inform., Vientiane, Laos.

Bedel, E., 1921: Buprestides. *In:* Fauna des Coléoptères du Bassin de la Seine, 4,2: 186–91.

Benoit, P., 1964: Comparative morphology of some *Chrysobothris* larvae (Coleoptera, Buprestidae) of Eastern Canada. – Can. Ent., 96: 1107–17.

– 1966: Description de la larve du *Melanophila acuminata* DeGeer et de quelques characters distinctifs du *Melanophila fulvoguttata* (Harris) (Coleoptera, Buprestidae). – Ibid., 98: 1208–11.

Bétis, L., 1914: In Caillol H.: Catalogue de Coléoptères du Provence. II. – Bull. Soc. et sc. arch. Drag., 28: 456–529.

Bílý, S., 1971: The larva of *Ptosima flavoguttata* (Ill.) (Coleoptera, Buprestidae). – Acta ent. bohemoslov., 68: 18–22.

– 1972: The larva of *Dicerca (Dicerca) berolinensis* (Herbst) (Coleoptera, Buprestidae) and the case of prothetely in this species. – Ibid., 69: 266–69.

– 1974: Zur Biologie einheimischer Käferfamilien. 13. Buprestidae. – Ent. Ber., 2: 67–79.

– 1975a: The larvae of eight species of the genus *Anthaxia* Eschscholtz, 1829 from Central Europe (Coleoptera, Buprestidae). – Acta ent. forest., 2: 63–82.

– 1975b: Larvae of European species of the genus *Chrysobothris* Eschsch. (Coleoptera, Buprestidae). – Acta ent. bohemoslov., 72: 418–24.

– 1977: Klíč k určování československých krasců (Coleoptera, Buprestidae). – Praha.

Bjørnstad, A. & Zachariassen, K. E., 1975: *Agrilus pratensis* Ratz. (Coleoptera, Buprestidae) new to Norway. – Norw. J. Ent., 22: 83–84.

Brahm, N. J., 1790. Insectenkalender für Sammler und Oekonomen. 1. – Mainz.

Burke, H. E., 1919; Biological notes on some flatheaded barkborers of the genus *Melanophila*. – J. econ. Ent., 12: 105–08.

Casey, T. L., 1909: Studies in the American Buprestidae. – Proc. Wash. Acad. Sci., 11, 2: 47–178.

Castelnau, F. L. L. & Gory, H. L., 1837 & 39: Histoire naturelle et iconographie des Coléoptères. 1 & 2. – Paris.

Chevrolat, C. A. A., 1837: Centurie de Buprestides. – Revue Ent. (Silbermann), 5: 41–110.

– 1838: Descriptions de trois *Buprestis* et d'un superbe *Cyphus* nouveau. – Revue zool., 1838: 55–56.

Cobos, A., 1980: Ensayo sobre los géneros de la subfamília Polyćestinae (Coleoptera, Buprestidae). Parte I. – Eos, 54: 15–94.

Creutzer, C., 1799: Entomologische Versuche. – Wien.

Curtis, J., 1825: British entomology, etc. 2 (2). – London.

– 1840: Descriptions of rare or interesting indigenous Insects. – Ann. nat. Hist., 5: 274–82.

Dahlgren, G., 1964: *Dicerca alni* Fisch., *berolinensis* Hbst. und *aenea* L. – Ent. Tidskr., 85: 203–04.

DeGeer, C., 1774: Mémoires pour servir à l' histoire des Insectes. 4. – Stockholm.

Dyke, E. C. van, 1939: An ancient beetle. – Pan-Pacif. Ent., 15: 154.

Eschscholtz, J. F., 1829–33: Zoologischer Atlas etc. – Berlin.

Fabricius, J. C., 1775: Systema Entomologiae. – Flensburgi et Lipsiae.

– 1777: Genera Insectorum. – Chilonii. Praefatio 1776. Editio 1777.

- 1781: Species Insectorum. - Hamburgi et Kilonii.
- 1787: Mantissa Insectorum. 2. - Hafniae.
- 1792: Entomologia Systematica. 1. - Hafniae.
- 1793: Entomologia Systematica Emendata et Aucta. 2. - Hafniae.
- 1794: Entomologica Systematica Emendata et Aucta. 4. - Hafniae.
- 1801: Systema Eleutheratorum. - Kiliae.

Fallén, C. F., 1802: Observationes entomologicae. Pars 1. - Dissert. Lund.

Fischer, W. G., 1823-24: Entomographia imperii russici. 2. - Mosquae.

Fourcroy, A. F., 1785: Entomologia parisiensis, etc. 2 Vols. - Paris.

Ganglbauer, L., 1889: Anleitung zum Käfersammeln in den Alpen. - Mitt. Sekt. Naturk. Österr. Tour.-Club, 1: 7-10.

Gebler, F. A., 1833: Notae et Additamenta ad Catalogum Coleopterorum Siberiae etc. - Bull. Soc. Nat. Moscou, 6: 262-309.

Germar, E. F., 1834: Fauna Insectorum Europeae. 3. - Halae.

Gernet, E. v., 1867: Eine Nachricht über Heuschrecken aus den 16. Jahrhundert. - Horae Soc. ent. Ross., 5: 157-160.

Gistl, J., 1834: Die Insecten-Doubletten aus der Sammlung des Grafen Rudolf von Jenison Walworth. - München.
- 1856: Die Mysterien der europäischen Insectenwelt. - Kempten.

Gory, H. L. & Laporte de Castelnau, F. L. L., 1841: Histoire naturelle et iconographie des insectes Coléoptères. IV. Suppl. Buprestides. Paris.

Goureau, C. C., 1843: Note pour servir à l' historie de l' *Agrilus biguttatus.* - Annls Soc. ent. F., (2) 1: 23-30.

Grill, C., 1896: Catalogus coleopterorum Scandinaviae, Daniae et Fenniae. 2 pts. - Stockholm .

Gyllenhal, L., 1808: Insecta Suecica descripta. 1. - Scaris.
- 1817: *In.* Schoenherr C. J.: Appendix ad Synonymiam Insectorum. 1. - Scaris.

Hansen, V., 1966: Smeldere og Pragtbiller. - Danm. Fauna, 74. København.

Harde, K. W., 1979: 38. Familie: Buprestidae (Prachtkäfer). *In.* Freude H., et al.: Die Käfer Mitteleuropas. Bd. 6 Krefeld.

Harris, T. W., 1829: Contribution to Entomology. No. VII. - New Engl. Farmer, 8: 2-3.

Hellrigl, K. G., 1972: Revision der westpaläarktischen Arten der Prachtkäfergattung *Lampra* Lac. (Coleoptera, Buprestidae). - Annln Naturh. Mus. Wien, 76: 649-708.
- 1978: Ökologie und Brutpflanzen europäischen Prachtkäfer (Coleoptera, Buprestidae). - Z. angew. Ent., 85: 167-91.

Herbst, J. F. W., 1780: Beschreibung aller Prachtkäfer, die bisher bei Berlin gefunden sind. - Schr. berl. Ges. naturf. Fr. Berl., 1: 85-100.
- 1787: Kurze Einleitung zur Kenntniss der Insecten für Ungeübte und Anfänger. 8. - Berlin u. Stralsund.
- 1801: Natursystem aller bekannten in- und ausländischen Insecten. 9. - Berlin.

Hering, E. M., 1951: Biology of the Leaf Miners. - Dr. W. Junk, 's-Gravenhage.

Hübner, J., 1816: Sammlung europäischer Schmetterlinge. 4. Noctuae, Eulen. - Augsburg.

Huggert, L., 1967: Några sällsyntare Coleoptera. - Ent. Tidskr., 88: 170-73.
- 1974: Anteckninger om Coleoptera. - Ibid., 95: 100-06.

Illiger, J. C. W., 1803: Magazin für Insectenkunde. 2. - Braunschweig.

Jacquelin du Val, C., 1859: Genera des Coléoptères d' Europe. ? - Paris.

Jacquin, N. J., 1778-81: Miscellanea austriaca ad botanicam, chemiam et historiam naturalem spectantia. 2 Vols. - Vindobonae.

Jodal, I., 1965: *Agrilus suvorovi populneus* Schaef., stetocina mladih topolovich zasada. - Topola, 9:

94

32–34.

Kangas, E., 1947: Studien über die *Agrilus*-larven (Coleoptera, Buprestidae). – Annls Ent. Fenn., 13: 104–13.

Kerremans, C., 1892: Catalogue synonymique des Buprestides décrits de 1758 à 1890. – Mém. Soc. ent. Belg., 1: 1–304.

– 1893: Essai de groupement des Buprestides. – Annls Soc. ent. Belg., 37: 94–122.

– 1900: Buprestides Indo-malais. III. – Mém. Soc. ent. Belg., 7: 61–93.

Kiesenwetter, H. v. 1857: Coleoptera, 1. *In* Erichson, W. F. Naturgeschichte der Insecten Deutschlands. 4. – Berlin.

Kirby, W., 1837: Entomologica Boreali-americana, etc. 4. – London.

Krogerus, R., 1922: Studien über *Agrilus*-Arten. – Notul. ent., 2: 109–113.

– 1925: Zur Kenntnis der Agriliden Finnlands. – Meddn Soc. Fauna Flora fenn., 49: 70–76.

Kurosawa, Y., 1959: A revision of the leaf-mining Buprestid-beetles from Japan and the Loo-Choo Islands. – Bull. nat. Sci. Mus. Tokyo, 43: 202–68.

Küster, H. C., 1844–46: Die Käfer Europas. – Nürnberg.

Lacordaire, J. T., 1835: *In* Boisduval J. B. A.; Fauna entomologue des environs de Paris, etc. – Paris.

– 1857: Histoire naturelle des insectes. Genera des Coléopteres, etc. 4. – Paris.

Latreille, P. A., 1810: Considérations générales, etc. – Paris.

Le Conte, J. L., 1852: General Remarks upon the Coleoptera of Lake Superior. *In* Agassiz L.: Lake Superior, 4. New York.

– 1857: Index to the Buprestidae of the United States, described in the work of Laporte and Gory. – Proc. Acad. nat. Sci. Philad., 4: 6–11.

Leller, I. E., 1947: *Agrilus biguttatus* Fabr. – Ent. Tidskr., 68: 13–15.

Lekič, M., 1959: Vrbin pistenar (*Agrilus acutangulus* Théry). Zaöt. Bilja, 54. 3–22.

Lewis, G., 1893: On the Buprestidae of Japan. – J. Linn. Soc., (Zool.), 24: 327–38.

Lindroth, C. H. (ed.), 1960: Catalogus Coleopterorum Fennoscandiae et Daniae. – Lund.

Linné, C., 1758: Systema naturae. Ed. 10, vol. 1. Holmlae.

1761: Fauna svecica. Ed. 2. Stockholmiae.

– 1767: Systema naturae. Ed. 12, reformata. 1 (2). – Stockholmiae.

Lundberg, S., 1957: Iakttagelser över skalbaggar från Norrbotten. – Ent. Tidskr., 77: 181–186.

1960: Bidrag till kännedomen om svenska Coleoptera. 3. – Ibid., 81: 108–12.

– 1961: Bidrag till kännedom om svenska Coleoptera. 4. – Ibid., 82: 64–68.

– 1962: Bidrag till kännedom om svenska skalbaggar. 5. Agrilus guerini Lac. – Ibid., 83: 172–74.

– 1963a: Bidrag till kännedomen om svenska Coleoptera. 6. – Ibid., 84: 119–24.

– 1963b: Bidrag till kännedom om svenska skalbaggar. 7. – Ibid., 84: 242–46.

– 1969: Bidrag till kännedom om svenska skalbaggar, 12. – Ibid., 90: 217–24.

– 1972: Bidrag till kännedom om svenska skalbaggar. 13. – Ibid., 93: 42–56.

– 1973: Bidrag till kännedom om svenska skalbaggar (Coleoptera). 14. – Ibid., 94: 28–33.

– 1975: Bidrag till kännedom om svenska skalbaggar (Coleoptera). 15. – Ibid., 96: 8–13.

– 1978: Skalbaggsarter som inte återfunnits i Sverige på lång tid-några tips (Coleoptera). – Ibid., 99: 121–126.

– 1980a: Fynd av för Sverige nya skalbaggsarter rapporterade under åren 1978–79. – Ibid., 101: 91–93.

– 1980b: Bidrag til kännedom om svenska skalbaggar. 19. – Ibid., 101: 95–97.

Lundberg, S., Baranowski, R., Nylander, U., 1971: Några interessanta skalbaggsfynd i Halltorps hage. – Ibid., 92: 286–89.

Lundblad, O., 1943: Några skalbaggsfynd. – Ibid., 64: 192.

Lysholm B., 1924: Coleoptera med nordgrænse i det Trondhjemske. – Norsk ent. Tidsskr., 1: 274–82.
– 1937: Coleopterfaunaen i Trøndelag. – Ibid., 4: 143–82.
Mannerheim, C. G., 1837: Enumération des Buprestides etc. – Bull. Soc. Nat. Moscou, 10: 3–126.
Marquet, C., 1869: Descriptions de Coléoptères nouveaux. – Abeille, 6: 368.
Marseul, S., 1865: Monographie des Buprestides d' Europe, du Nord de l'Afrique et de l'Asie. – Abeille, 2: 1–396.
– 1869: Descriptions de Coléoptères nouveaux. – Ibid., 6: 379–89.
Ménétries, E., 1832: Catalogue raisonné des objets de Zoologie recueillis dans un voyage au Caucase et jusqu' aux frontières actuelles de la Perse. – St. Petersbourg.
Méquignon, A., 1907: Synonymies de Coléoptères paléarctiques. – Bull. Soc. ent. Fr., 1907: 119–20.
Morgan, D. F., 1966: The biology and behaviour of the beech Buprestid Nascioides enysi (Sharp.), (Coleoptera, Buprestidae) with notes on its ecology and possibilities for its control. – Trans. R. Soc. N. Z., Zoology, 7 (11): 159–170.
Motschulsky, V., 1859: Coléoptères du Gouvernement de Jakoutsk. – Mélang. biol. Bull. Acad. Sci. St. Petersb., 3: 221–238.
Mulsant, E., 1863: Descriptions de quelques Coléoptères nouveaux ou peu connus. – Annls Soc. linn. Lyon, 10: 4–29.
Munster, T., 1921: Bidrag til Norges koleopterfauna. – Norsk ent. Tidsskr., 1: 87–100.
– 1922: Tillæg til Norges koleopterfauna. – Ibid., 1: 118–135.
– 1927: Tillæg og bemærkninger til Norges koleopterfauna. – Ibid., 2: 158–200.
Nilsson, A. & Andersen, J., 1977: Finds of Coleoptera from Northern Norway. – Norw. J. Ent., 24: 7–9.
Obenberger, J., 1912: Neue palaearktische Buprestiden. – Koleopt. Rdsch., 1: 65–70.
– 1917: Studien über paläarktischen Buprestiden. II. – Wien. ent. Ztg, 36: 209–218.
– 1924: Symbolae and speciarum regionis palaearcticae Buprestidarum cognitionem. – Jub. Sbor. čsl. Spol. ent., 1924: 6–59.
– 1927: De novis Buprestidarum regionis palaearcticae speciebus. IX. – Cas. čsl. Spol. ent., 24: 15–20.
– 1930: Buprestidarum supplementa palaearctica. VI. – Ibid., 27: 102–115.
– 1936: Fam. Buprestidae V. In Junk W. & Schenkling S.: Coleopterorum Catalogus. Pars 152. – Berlin.
– 1955: Matériaux pour servir a la connaissance des Bupréstides paléarctiques. I. – Acta ent. Mus. Nat. Pragae, 30: 41–47.
Olivier, G. A., 1790: Entomologie, ou histoire naturelle des Insectes, etc. Coléoptères, 2. – Paris.
Pallas, P. S., 1782: Icones Insectorum, etc. 2. – Erlangae.
Palm, T., 1962: Zur Kenntnis der früheren Entwicklungsstadien schwedischer Käfer. 2. Buprestiden-larven, die in Bäumen leben. – Opusc. Ent., 27: 65–78.
Panzer, G. W. F., 1789: Beschreibung seltner Insekten. – Der Naturforscher, St. 24: 1–35.
Paykull, G., 1799: Fauna Suecica: Insecta. 2. – Upsaliae et Hafniae.
Perris, E., 1854: Histoire des Insectes du Pin maritime. – Annls Soc. ent. Fr., (3) 2: 85–160, 593–646.
– 1877: Larves de Coléoptères. – Paris.
Pic, M., 1918: Notes diverses, descriptions et diagnoses. Échange, 34: 1–3, 9–11, 13–15, 17–19, 21–24.
– 1922: Notes diverses, descriptions et diagnoses. – Ibid., 38: 29–30.
Piller, M. & Mitterpacher, L., 1783: Iter per Poseganam Slavoniae provinciam mensibus Junio et

Julio 1782 susceptum. – Budae.

Plochich, V. S., 1969: K biologii i ekologii temnoi zlatki *Agrilus ater* (L.) (Col., Buprestidae) – stvolovogo vreditela topolei. – Ent. Obozr., 48: 493–99.

Pochon, H., 1963: Buprestiden-Ausbeute aus Spanien (Catalonien) und Neubeschreibung zweier "*Agrilus*". – Misc. zool., 1: 65–70.

– 1964: Coleoptera – Buprestidae. Insecta Helvetica, 2. – Lausanne.

Ponza, L., 1805: Coleoptera Salutiensia, etc. – Mém. Acad. Sci. Turin, 14: 29–94.

Ragusa, E., 1893: Catalogo regionato dei Coleotteri di Sicilia – Buprestidae. – Natural. Sicil., 12.

Ratzeburg, J. T. C., 1837: Die Forstinsekten etc. 1. ed. – Berlin.

– 1839: Die Forstinsekten etc. 2. ed. – Berlin.

Redtenbacher, L., 1849: Fauna Austriaca. Die Käfer, nach der analytischen Methode bearbeitet. – Wien.

Reitter, E., 1911: Die Käfer des Deutschen Reichs. – Fauna Germanica, 3. – Stuttgart.

Rey, C., 1891: Remarques en passant. – Échange, 7: 4–5.

Richter, A. A., 1944: Zlatki *Anthaxia* Kavkaza (Col., Buprestidae). – Zool. Sb. Erevan, 3: 109–30.

– 1949: Zlatki (Buprestidae) I. – Fauna SSSR, Tom 13, vyp. 2. – Moskva-Leningrad.

– 1952: Zlatki (Buprestidae) II. – Fauna SSSR, Tom 13, vyp. 4. – Moskva-Leningrad.

Rivney, E., 1946: Ecological and physiological studies on *Capnodis* spp. (Col., Buprestidae) in Palestine. III. Studies on the adult. – Bull. ent. Res., 37: 273–80.

Rosenhauer, W. G., 1856: Die Thiere Andalusiens etc. – Erlangen.

Rossi, P., 1790: Fauna Etrusca, etc. 2 Vols. – Liburni.

Saalas, U., 1923: Die Fichtenkäfer Finnlands. II. – Helsinki.

Sahlberg, J., 1900: Catalogus Coleopterorum Faunae fennicae geographicus – Acta Soc. Fauna Flora Fenn., 19 (4): 1–132.

Saunders, E., 1871: Catalogus Buprestidarum synonymicus et systematicus. London.

Say, T., 1823–24. Descriptions of Coleopterous Insects collected in the late expedition to the Rocky Mountains. – Journ. Acad. nat. Sci. Philad., 3 (1823). 139–216, (1824) 238–82; 298–331, 403–62.

Schaefer, L., 1946: Formes nouvelles de Buprestides – Bull. Soc. linn. Lyon, 16: 71–73.

– 1949: Les Buprestides de France. – Paris.

Schäffer, J. C., 1766: Icones Insectorum etc. Pt. 1. – Regensburg.

Schigdte, J. C., 1870: De Metamorphosi Eleutheratorum Observationes. – Naturh. Tidsskr., 6: 353–378.

Schönherr, C. J., 1817: Synonyma Insectorum, etc., Bd. 1 (3). – Scaris.

Schrank, F. P., 1781: Enumeratio Insectorum Austriae indigenorum. – Klett.

– 1789: Entomologischen Beobachtungen. – Der Naturforscher St. 24: 60–90.

Schreibers, C. F. A., 1843: *In* Sturm J.: Catalog der Käfer-Sammlung von J. Sturm. – Nürnberg.

Scopoli, J. A., 1763: Entomologia Carniolica, etc. – Vindobonae.

Semenov-Tianschanskij, A. P., 1895: Coleoptera nova Rossiae Europaeae Caucasique. – Trudy russk. ent. Obshch., 29: 242–250.

– 1935: Analecta coleopterologica. – Ent. Obozr., 25: 271–281.

– & Richter, A. A., 1934: Notes sur les Chrysobothris peu connus de l' Asie palearctique et description de trois especes nouvelles. – Bull. Soc. ent. Fr.: 94.

Siebke, H., & Sparre-Schneider, J., 1875: Catalogum Coleopterorum Continens. *In* Enumeratio Insectorum Norvegicum, 2: 61–334. Christiania.

Silfverberg, H., 1977: Nomenclatoric Notes on Coleoptera Polyphaga. – Notul. ent., 57: 91–94.

– 1979: Enumeratio Coleopterorum Fennoscandiae et Daniae. – Helsingfors.

Soldatova, E. A., 1969: Morfologiceskaia charakteristika licinok zlatok roda *Melanophila* Eschsch.

evropeiskoi casti SSSR. – Nauc. dokl. vys. skoly, Biol. nauki, 1: 12–16.
- 1970: Taksonomiceskoe znacenie morfologiceskich priznakov licinok zlatok rodov *Anthaxia* i *Cratomerus*. – Zool. Zh., 49: 61–71.
- 1973: Osobennosti stroenia zheludka licinok zlatok otnosiascichsia k tribu Anthaxiini, Kisanthobiini i Melanophiliini (Coleoptera, Buprestidae). – Ént. Obozr., 52: 582–585.
Solier, A. J. J., 1833: Essais sur les Buprestides. – Annls Soc. ent. Fr. 2: 261–316.
Stepanov, V. N., 1953: Dva novych dlja fauny SSSR vida uzkotelych zlatok roda *Agrilus* Curt. iz podroda *Epinagrilus* V. Step. subgen. nov. – Zool. Zh., 33: 114–119.
Stephens, J. F.., 1829: A systematical catalogue of British Insects, etc. – London.
- 1830: Illustrations of British entomology, etc. 3. – London.
Stierlin, W. G., 1868: Beschreibung zweier neuer Käferarten. – Mitt. schweiz. ent. Ges., 2: 345–47.
Strand, A., 1938: Bemerkninger vedkommende koleopterfaunaen i Rana. – Norsk ent. Tidsskr., 5: 83–86.
- 1943: Inndeling av Norge til bruk ved faunistiske oppgaver. – Ibid., 6: 208–224.
- 1946: Nord-Norges Coleoptera. – Tromsø Mus. Årsh., 67: 1–629.
- 1957: Koleopterologiske bidrag VIII. – Norsk ent. Tidsskr., 10: 110–118.
- 1958: Koleopterologiske bidrag IX. – Ibid., 10: 189–194.
- 1962: *Anthaxia quadripunctata* L. og *godeti* Cast. et Gory (*submontana* Obnb.) (Col., Buprestidae). – Ibid., 12: 36–38.
- 1965: Koleopterologiske bidrag XI. – Ibid., 13: 82–91.
- 1970: Additions and corrections to the Norwegian part of Catalogus Coleopterorum Fennoscandiae et Daniae. – Ibid., 17: 125–145.
- 1977: Additions and corrections to the Norwegian part of Catalogus Coleopterorum Fennoscandiae et Daniae. Second series. – Norw. J. Ent., 24: 159–165.
Strand, A. & Hanssen, H. K., 1932: Målselvens Koleoptera. – Norsk ent. Tidsskr., 3: 17–71.
Sulzer, J. H., 1776: Abgekürzte Geschichte etc. 2 pts. – Winterthur.
Théry, A., 1942: Coléopteres Buprestides. Faune de France, 41. – Paris.
Thomson, C. G., 1864: Skandinaviens Coleoptera, synoptisk bearbetade. 6. – Lund.
Thunberg, C. P., 1784: Dissertatio entomologica novas insectorum species sistens. Pars 4. – Upsaliae.
Villers, C. J., 1789: Caroli Linnaei entomologia. 1. – Lugduni.
Voet, J. E., 1806: Catalogus systematicus Coleopterorum. 2. – La Haye.
Volkovič, M. G., 1979: K morfologii licinok zlatok roda *Acmaeoderella* Cobos (Coleoptera, Buprestidae). – Trudy zool. Inst. Leningr., 83: 21–38.
Weiss, H. B., 1914: *Agrilus politus* Say infesting Roses. – J. econ. Ent., 7: 438–440.
Yano, T., 1952: The developmental stages of two genera of Trachyinae, *Trachys* and *Habroloma*, of Shikoku, Japan (Coleoptera, Buprestidae). – Trans. Shikoku ent. Soc., 3: 17–40.
Zachariassen, K. E., 1972: Notes on distribution of Coleoptera in Norway. – Norsk ent. Tidsskr., 19: 169–170.
- 1977: New finds of Coleoptera in Norway. – Norw. J. Ent., 24: 147–48.
- 1979: New records of Coleoptera in Norway. – Fauna Norw., Ser. B, 26: 5–7.

Catalogue

	№	Germany	G. Britain	SJ	EJ	WJ	NWJ	NEJ	F	LFM	SZ	NWZ	NEZ	B	Sk.	Bl.
Buprestis splendens F.	1										i					
B. r. rustica L.	2	●	●							i		i				●
B. h. haemorrhoidalis Herbst	3	●			●					i		i	i		●	●
B. o. octoguttata L.	4	●														
B. n. novemmaculata L.	5	●				i			●						●	●
Dicerca moesta (F.)	6	●											î			
D. f. furcata Thbg.	7	●														●
D. a. aenea (L.)	8	●														●
D. alni (Fischer)	9	●														●
Poecilonota v. variolosa (Payk.)	10	●													●	●
Scintillatrix rutilans (F.)	11	●														
Melanophila acuminata (DeGeer)	12	●	●		●					i			î	●	●	●
Phaenops cyanea (F.)	13	●	●									î			●	●
Anthaxia n. nitidula (L.)	14	●	●													
Anthaxia manca (L.)																
A. quadripunctata (L.)	15	●	●	●									●		●	●
A. godeti Gory & Cast.	16	●														
A. similis Saunders	17	●														
A. h. helvetica Stierlin																
Chalcophora m. mariana (L.)	18	●		i									i		●	
Chrysobothris a. affinis	19	●								●	●			●	●	●
C. c. chrysostigma (L.)	20	●	●											●	●	
Agrilus angustulus Ill.	21	●	●	●	●	●	●		●	●	●	●	●	●	●	●
A. a. ater (L.)	22	●														
A. a. aurichalceus Redt.	23	●														●
A. a. paludicola Krog.	24															
A. b. betuleti (Ratz.)	25	●													●	●
A. biguttatus (F.)	26	●	●												●	●
A. c. convexicollis Redt.	27	●														
A. cyanescens Ratz.	28	●				●	●	●								
A. guerini Lac.	29	●														
A. hyperici Creutzer																
A. integerrimus (Ratz.)	30	●														
A. laticornis Ill.	31	●	●	●	●				●	●			●		●	●
A. mendax Manneil.	32	●														
A. olivicolor Kiesenw.	33	●													●	●
A. p. pratensis (Ratz.)	34	●	●												●	●
A. p. pseudocyaneus Kiesenw.	35	●														

SWEDEN

	Hall.	Sm.	Öl.	Gtl.	G. Sand.	Ög.	Vg.	Boh.	Dlsl.	Nrk.	Sdm.	Upl.	Vstm.	Vrm.	Dlr.	Gstr.	Hls.	Med.	Hrj.	Jmt.	Ång.	Vb.	Nb.	Ås. Lpm.	Ly. Lpm.	P. Lpm.	Lu. Lpm.	T. Lpm.
1										●		●																
2	●	●		●		●	●	●	●	●	●	●	●	●	●	●	●	●	●	●	●	●	●		●	●	●	
3		●	●		●	●	●	●	●	●	●	●	●	●			●			●			●					
4		●	●	●	●	●	●		●		●	●	●	●														
5		●		●		●	●				●	●																
6		●	●			●	●		●		●	●		●	●		●				●		●					
7		●	●			●	●			●	●		●	●	●		●		●		●	●			●	●	●	
8																												
9		●	●			●				●	●	●			●													
10	●	●	●		●	●	●	●	●	●	●	●	●								●	●						
11																												
12	●	●	●	●	●		●	●			●	●	●	●	●	●	●		●	●		●	●	●	●	●	●	
13		●	●	●	●	●	●			●	●	●	●	●	●	●	●		●			●	●	●	●			
14	S	W	E	D	E	N																						
15	●	●	●	●		●	●	●	●	●	●	●	●	●	●	●	●	●	●	●	●	●	●	●	●	●		
16		●	●								●																	
17		●	●	●		●				●	●				●													
18		●	●	●		●				●	●				●													
19	●	●	●		●	●	●			●	●																	
20	●	●	●		●	●	●	●	●	●	●	●	●	●	●	●	●	●			●	●	●	●	●	●	●	
21	●	●	●	●		●	●	●	●	●																		
22																												
23																												
24																	●			●	●					●	●	
25		●	●			●	●	●		●	●	●	●	●	●	●												
26		●	●			●	●			●																		
27			●	●																								
28																												
29		●																										
30																												
31	●	●	●	●		●	●	●			●		●															
32													●															
33		●	●			●	●				●																	
34		●	●			●	●		●	●	●	●		●		●												
35																												

		Ø+AK	HE (s+n)	O (s+n)	B (ø+v)	VE	TE (y+i)	AA (y+i)	VA (y+i)	R (y+i)	HO (y+i)	SF (y+i)	MR (y+i)	ST (y+i)	NT (y+i)	Ns (y+i)
Buprestis splendens F.	1															
B. r. rustica L.	2	●	●	●	●	●	●	●	●	●	●					
B. h. haemorrhoidalis Herbst	3	●	●		●		●	●								
B. o. octoguttata L.	4	●	●	?	●			●	●							
B. n. novemmaculata L.	5				?											
Dicerca moesta (F.)	6	●					●									
D. f. furcata Thbg.	7	●			●		●									
D. a. aenea (L.)	8	●			?											
D. alni (Fischer)	9															
Poecilonota v. variolosa (Payk.)	10	●	●		●		●	●								
Scintillatrix rutilans (F.)	11	●			●											
Melanophila acuminata (DeGeer)	12	●	●		●			●								●
Phaenops cyanea (F.)	13	●	●		●	●	●	●								
Anthaxia n. nitidula (L.)	14															
Anthaxia manca (L.)																
A. quadripunctata (L.)	15	●	●	●	●	●	●	●	●	●				●	●	●
A. godeti Gory & Cast.	16	●	●	●			●	●	●							
A. similis Saunders	17	●					●	●	●							
A. h. helvetica Stierlin																
Chalcophora m. mariana (L.)	18	●						●	●							
Chrysobothris a. affinis	19						●	●	●	●						
C. c. chrysostigma (L.)	20	●	●	●	●	●	●	●	●							
Agrilus angustulus Ill.	21	●					●	●	●	●						
A. a. ater (L.)	22															
A. a. aurichalceus Redt.	23															
A. a. paludicola Krog.	24		●													
A. b. betuleti (Ratz.)	25			●												
A. biguttatus (F.)	26					●		●								
A. c. convexicollis Redt.	27															
A. cyanescens Ratz.	28															
A. guerini Lac.	29															
A. hyperici Creutzer																
A. integerrimus (Ratz.)	30															
A. laticornis Ill.	31						●	●								
A. mendax Mannerh.	32															
A. olivicolor Kiesenw.	33	●			●											
A. p. pratensis (Ratz.)	34						●									
A. p. pseudocyaneus Kiesenw.	35															

102

	Nn (ø+v)	TR (y+i)	F (v+i)	F (n+ø)	Al	Ab	N	Ka	St	Ta	Sa	Oa	Tb	Sb	Kb	Om	Ok	ObS	ObN	Ks	LkW	LkE	Le	Li	Vib	Kr	Lr
														FINLAND											USSR		
1																									●		
2					●	●	●	●	●	●	●	●	●	●	●	●	●	●	●	●					●	●	
3					●	●	●	●	●	●	●	●	●	●	●			●	●	●					●	●	
4					●	●		●				●													●	●	
5					●	●			●	●			●												●	●	
6					●	●					●	●	●				●								●	●	
7					●	●			●	●	●	●	●	●			●	●		●					●	●	
8																											
9	●	●							●	●		●		●											●	●	
10					●	●			●	●	●		●												●	●	
11						●					●																
12	●	●	●	●	●	●			●	●	●		●	●	●	●	●	●	●	●	●	●	●	●	●	●	●
13					●	●			●	●	●														●	●	
14																											
15					●	●	●	●	●	●	●	●	●	●	●	●	●	●	●	●		●			●	●	●
16																											
17																											
18					●	●			●	●	●														●	●	
19																											
20					●	●	●	●	●	●	●	●	●	●	●	●	●	●	●	●					●	●	
21					●	●																					
22					●	●																					
23											●																
24					●	●	●		●		●		●	●	●		●	●		●					●	●	●
25					●	●	●	●	●		●														●	●	
26																											
27																											
28																											
29																											
30					●				●	●																●	
31						●																					
32					●	●		●	●	●															●	●	
33																									●		
34					●	●	●		●	●		●														●	●
35											●														●	●	●

		Germany	G. Britain	SJ	EJ	WJ	NWJ	NEJ	F	LFM	SZ	NWZ	NEZ	B	Sk.	Bl.
A. subauratus (Gebl.)	36	●														
A. sulcicollis Lac.	37	●									●	●	●		●	●
A. suvorovi populneus Schaef.	38	●														●
A. v. viridis (L.)	39	●	●	●	●	●	●		●		●	●	●	●	●	●
Aphanisticus pusillus (Oliv.)	40	●	●	●	●	●			●	●	●	●	●	●	●	●
A. emarginatus (Oliv.)																
Trachys m. minutes (L.)	41	●	●	●	●	●					●				●	●
T. troglodytes Gyll.	42	●	●	●							●		●			
T. scrobiculatus Kiesenw.	43	●	●								●					
Habroloma geranii (Silfverbg.)	44	●	●		●										●	●

		Ø + AK	HE (s+n)	O (s+n)	B (ø+v)	VE	TE (y+i)	AA (y+i)	VA (y+i)	R (y+i)	HO (y+i)	SF (y+i)	MR (y+i)	ST (y+i)	NT (y+i)	Ns (y+i)
A. subauratus (Gebl.)	36															
A. sulcicollis Lac.	37	●					●	●	●	●						
A. suvorovi populneus Schaef.	38						●	●								
A. v. viridis (L.)	39	●	●	●	●	●	●	●	●	●				●	●	●	●
Aphanisticus pusillus (Oliv.)	40															
A. emarginatus (Oliv.)																
Trachys m. minutes (L.)	41	●	●	●	●	●	●	●	●	●		●	●			
T. troglodytes Gyll.	42															
T. scrobiculatus Kiesenw.	43	●					●									
Habroloma geranii (Silfverbg.)	44	●		●	●	●	●	●								

	Hall.	Sm.	Öl.	Gtl.	G. Sand.	Üg.	Vg.	Boh.	Dlsl.	Nrk.	Sdm.	Upl.	Vstm.	Vrm.	Dlr.	Gstr.	Hls.	Mcd.	Hrj.	Jmt.	Ång.	Vb.	Nb.	Ås. Lpm.	Ly. Lpm.	P. Lpm.	Lu. Lpm.	T. Lpm.
36		●				●					●	●																
37	●	●	●			●			●		●	●	●			●												
38		●				●			●	●	●	●																
39	●	●	●	●		●	●	●	●	●	●	●	●	●	●	●	●	●		●	●	●	●	●	●	●	●	●
40	●	●	●			●	●	●																				
41	●	●	●	●		●	●	●	●	●	●	●	●	●	●	●	●			●	●	●	●				●	●
42	●	●	●			●	●		●																			
43																												
44		●	●	●		●	●	●	●	●	●	●	●	●	●	●				●								

	Nn (ø+v)	TR (y+i)	F (v+i)	F (n+ø)	Al	Ab	N	Ka	St	Ta	Sa	Öa	Tb	Sb	Kb	Om	Ok	ObS	ObN	Ks	LkW	LkE	Le	Li	Vib	Kr	Lr
36						●	●	●	●	●	●				●										●	●	
37						●	●			●	●														●		
38																											
39	●	●			●	●	●	●	●	●	●	●	●	●	●	●	●	●	●	●	●				●	●	
40																											
41						●	●	●	●	●	●	●	●	●	●	●	●								●	●	
42																											
43																											
44					●	●	●		●	●	●		●		●		●								●	●	

Index

Synonyms are given in italics. The number in bold refers to the main treatment of the taxon.

Author's address:
Dr. Svatopluk Bílý
Dept. of Entomology
National Museum
Kunratice 1, 148 00 Praha 4
Czechoslovakia

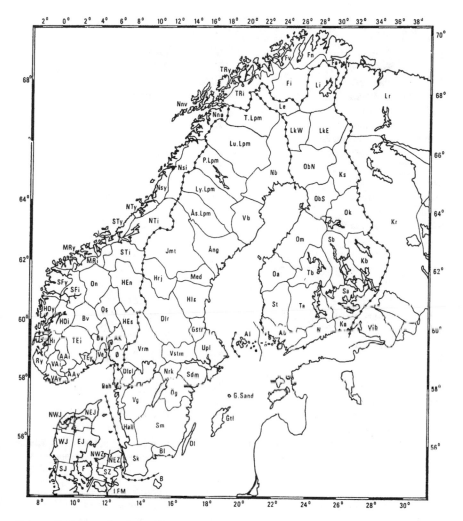

List of abbreviations for the provinces used throughout the text, on the map and in the following tables.

DENMARK

SJ	South Jutland	LFM	Lolland, Falster, Møn
EJ	East Jutland	SZ	South Zealand
WJ	West Jutland	NWZ	North West Zealand
NWJ	North West Jutland	NEZ	North East Zealand
NEJ	North East Jutland	B	Bornholm
F	Funen		

SWEDEN

Sk.	Skåne	Vrm.	Värmland
Bl.	Blekinge	Dlr.	Dalarna
Hall.	Halland	Gstr.	Gästrikland
Sm.	Småland	Hls.	Hälsingland
Öl.	Öland	Med.	Medelpad
Gtl.	Gotland	Hrj.	Härjedalen
G. Sand.	Gotska Sandön	Jmt.	Jämtland
Ög.	Östergötland	Ång.	Ångermanland
Vg.	Västergötland	Vb.	Västerbotten
Boh.	Bohuslän	Nb.	Norrbotten
Dlsl.	Dalsland	Ås. Lpm.	Åsele Lappmark
Nrk.	Närke	Ly. Lpm.	Lycksele Lappmark
Sdm.	Södermanland	P. Lpm.	Pite Lappmark
Upl.	Uppland	Lu. Lpm.	Lule Lappmark
Vstm.	Västmanland	T. Lpm.	Torne Lappmark

NORWAY

Ø	Østfold	HO	Hordaland
AK	Akershus	SF	Sogn og Fjordane
HE	Hedmark	MR	Møre og Romsdal
O	Opland	ST	Sør-Trøndelag
B	Buskerud	NT	Nord-Trøndelag
VE	Vestfold	Ns	southern Nordland
TE	Telemark	Nn	northern Nordland
AA	Aust-Agder	TR	Troms
VA	Vest-Agder	F	Finnmark
R	Rogaland		

n northern   s southern   ø eastern   v western   y outer   i inner

FINLAND

Al	Alandia	Kb	Karelia borealis
Ab	Regio aboensis	Om	Ostrobottnia media
N	Nylandia	Ok	Ostrobottnia kajanensis
Ka	Karelia australis	ObS	Ostrobottnia borealis, S part
St	Satakunta	ObN	Ostrobottnia borealis, N part
Ta	Tavastia australis	Ks	Kuusamo
Sa	Savonia australis	LkW	Lapponia kemensis, W part
Oa	Ostrobottnia australis	LkE	Lapponia kemensis, E part
Tb	Tavastia borealis	Li	Lapponia inarensis
Sb	Savonia borealis	Le	Lapponia enontekiensis

USSR

Vib   Regio Viburgensis   Kr   Karelia rossica   Lr   Lapponia rossica

Printed in the United States
By Bookmasters